Heritage, Conservation and Communities

Public participation and local community involvement have taken centre stage in heritage practice in recent decades. In contrast with this established position in wider heritage work, public engagement with conservation practice is less well developed. The focus here is on conservation as the practical care of material cultural heritage, with all its associated significance for local people. How can we be more successful in building capacity for local ownership and leadership of heritage conservation projects, as well as improving participative involvement in decisions and in practice?

This book presents current research and practice in community-led conservation. It illustrates that outcomes of locally-led, active participation show demonstrable social, educational and personal benefits for participants. Bringing together UK and international case studies, the book combines analysis of theoretical and applied approaches, exploring the lived experiences of conservation projects in and with different communities. Responding to the need for deeper understanding of the outcomes of heritage conservation, it examines the engagement of local people and communities beyond the expert and specialist domain.

Highlighting the advances in this important aspect of contemporary heritage practice, this book is a key resource for practitioners in heritage studies, conservation and heritage management. It is also relevant for the practising professional, student or university researcher in an emerging field that overarches professional and academic practice.

Gill Chitty is Director of the Conservation Studies programme and Centre for Conservation Studies in the Department of Archaeology, University of York. She was Head of Conservation at the Council for British Archaeology from 2005–12 and her professional experience in heritage conservation and public archaeology has been in local government, English Heritage (now Historic England) and as a consultant. Her doctoral research explored the influence of John Ruskin's work in shaping British conservation practice and her research interests continue around the political economy of heritage.

Heritage, Culture and Identity
Series editor: Brian Graham,
University of Ulster, UK

This series explores all notions of heritage – including social and cultural herit-age, the meanings of place and identity, multiculturalism, management and plan-ning, tourism, conservation and the built environment – at all scales from the global to the local. Although primarily geographical in orientation, it is open to other disciplines such as anthropology, history, cultural studies, planning, tour-ism, architecture/conservation and local governance and cultural economics.

For a full list of titles in this series, please visit www.routledge.com/Heritage-Culture-and-Identity/book-series/ASHSER-1231

Heritage, Conservation and Communities

Engagement, participation and capacity building

Edited by Gill Chitty

Routledge
Taylor & Francis Group

LONDON AND NEW YORK

First published 2017
by Routledge

2 Park Square, Milton Park, Abingdon, Oxfordshire OX14 4RN
711 Third Avenue, New York, NY 10017

Routledge is an imprint of the Taylor & Francis Group, an informa business

First issued in paperback 2018

British Library Cataloguing in Publication Data
A catalogue record for this book is available from the British Library

Library of Congress Cataloging in Publication Data
A catalog record for this book has been requested

ISBN: 978-1-4724-6800-0 (hbk)
ISBN: 978-1-138-33941-5 (pbk)

Typeset in Times New Roman
by Cenveo Publisher Services

Contents

Figures

Tables

Contributors

Ian Bapty (MA Cantab) is an archaeologist with over 20 years' experience of volunteer and community-based heritage and conservation projects. He worked as Project Officer and Manager with the British Trust for Conservation Volunteers (1992–9), as the English Heritage (now Historic England)/Cadw-funded Offa's Dyke Archaeological Management Officer (1999–2006) and as Senior Project Archaeologist with Herefordshire Council (2006–12). Ian's former role as Industrial Heritage Support Officer, based with the Ironbridge Gorge Museum Trust, supported the care of publicly accessible industrial heritage sites in England. He now works on the care of historic churches in Durham and Newcastle.

Lianne Brigham and Richard Brigham are founding administrators for the Facebook group York Past and Present. The group has grown in popularity since beginning in the winter of 2013 and currently has 8,500 members. York Past and Present runs a number of projects which involve its members in York's history and heritage. A recent project has been in collaboration with York's Mansion House, where members have photographed and packed the collections prior to a major development and redisplay.

Peter Brown was Director of York Civic Trust until his retirement in July 2015. Awarded an MBE for 'helping to preserve the heritage of York' in the Queen's New Year Honours list 2011, Peter came to York in the early 1980s to set up Fairfax House as an eighteenth-century house museum for the York Civic Trust, which opened to the public. He has established an international reputation for interpretation of the house museum and its collections. With books covering subjects ranging from the Noel Terry Collection of Furniture and Clocks to a ground-breaking millennium exhibition catalogue entitled *Eat, Drink and Be Merry*, he has also curated 25 exhibitions, notably *Views of York* (2012) and *In Praise of Hot Liquors* (1996). Peter is currently working on the publication of a new book, 'YORK 1660–1860'.

Gill Chitty is Director of the Conservation Studies programme and Centre for Conservation Studies in the Department of Archaeology, University of York. She was Head of Conservation at the Council for British Archaeology from 2005–12; her professional experience in heritage conservation and public archaeology has been in local government, with English Heritage (now Historic England) and as a consultant. Her doctoral research explored the influence of John Ruskin's work in shaping British conservation practice and her research interests continue around the political economy of heritage.

Sarah Court works with the International Centre for the Study of the Preservation and Restoration of Cultural Property (ICCROM) on their programme on Promoting People-Centred Approaches to Conservation. Based in Italy, she has worked for the Herculaneum Conservation Project for over a decade, promoting engagement initiatives for the local and international communities. She has also provided consultancy for heritage sites in countries such as Sri Lanka, Iraq and Cyprus, while providing capacity building support to organisations such as the Getty Conservation Institute. She has an MA in Roman archaeology and an MSc in heritage interpretation.

Alison Drake is the Chair of Castleford Heritage Trust. She was awarded an MBE in 2012 for services to her community and has been the driving force behind a number of regeneration projects in Castleford. These include bringing Channel 4 to the town for the Castleford Project, which helped draw in significant investment to improve the town centre and build an award-winning bridge over the River Aire. The Trust's current major project is Queen's Mill, Castleford.

Keith Emerick is a Research Associate at the University of York. He is a practising cultural heritage manager based in the north of England, currently working for Historic England (the national conservation and heritage agency) as an Inspector of Ancient Monuments. The role includes working and advising on ruins, archaeological sites and battlefields, covering the Palaeolithic to the Cold War. His principal research and work interests are in community heritage and the reconstruction and rehabilitation of ruins. Keith is the author of *Conserving and Managing Ancient Monuments: Heritage, Democracy and Inclusion* (Boydell and University of Newcastle 2014).

Lucía Gómez-Robles (DPhil, University of Granada) took degrees in Architecture in Grenada and Art History in Madrid. She has worked as an independent consultant in survey, documentation, conservation and restoration projects in different national and international contexts. She is currently Projects Director at DIADRASIS and Subdirector of Information and Communication at the National Coordination for the Conservation of the Cultural Heritage, National Institute of Anthropology and History (INAH) in México.

Helen Graham is University Research Fellow in Tangible and Intangible Heritage and Director, Centre for Critical Studies in Museums, Galleries and Heritage, University of Leeds, School of Fine Art, History of Art and Cultural Studies. Helen's research and teaching interests directly flow from working in learning and access teams in museums and coordinating community heritage projects concerned with the co-production of knowledge, archives and exhibits. Helen has recently acted as Principal Investigator on an Arts and Humanities Research Council 'Connected Communities' research project, 'How should decisions about heritage be made?', which explored 'how to increase participation from where you are'.

Joanne Harrison is an architect who, after several years in practice, returned to academia to specialise in heritage conservation. Having completed an MA at the University of York, she is now undertaking a PhD. Her research interests are in housing and regeneration, notably the nineteenth- and early twentieth-century terraces designed for the working classes. Current research is concerned with the heritage significance and protection of the back-to-back terraces in Leeds, and potential conflict with the viability of their continued use. The multi-disciplinary approach – spanning architectural, archaeological, historical and sociological methods – has a special focus on engaging community stakeholders.

Stella Jackson works part time for the Society for the Protection of Ancient Buildings as Lincolnshire Project Officer within the Maintenance Co-operatives Project. Prior to this she worked for English Heritage (now Historic England) in the Designation, Heritage at Risk and Government Advice teams. Stella's PhD research at the University of York looks at the contested nature of heritage.

Jukka Jokilehto is Special Advisor to the Director General of ICCROM, Professor at the University of Nova Gorica, Slovenia, and Visiting Professor at the University of York. His career at ICCROM (where he was Assistant Director General), on the World Heritage Committee and the International Council on Monuments and Sites (ICOMOS) International Training Committee has engaged him in international missions on cultural heritage in many parts of the world, including development of conservation master plans (Baku, Azerbaijan), management plans (Bam, Iran; Mtskheta, Georgia) and as an advisor on nominations to theWorld Heritage List (China, Japan, India, Iran, Italy, Ireland, Norway). His seminal work, *A History of Architectural Conservation* (Butterworth 1999), is currently in revision for a new edition.

Danai Koutromanou is a conservator specialising in conservation of archaeological material. Her PhD in Conservation Studies at the University of York researches public engagement in cultural heritage conservation. She has worked on the conservation of archaeological sites, museum collections and historic buildings, where she developed an interest in architectural conservation, taking an MA in Conservation of Historic Buildings at York in 2011; her dissertation was awarded the Duncan Gillard Memorial Medal. Danai is currently working on the excavations of the 26th Ephorate of Prehistoric and Classical Antiquities at the Faliron Delta in Athens.

Philip Leverton is a tutor at the University College of Estate Management, Reading, UK. He holds academic and professional qualifications in surveying, town planning and heritage management. His career has covered practice, teaching and research relating to planning, development, conservation and heritage management in the higher education, local government and NGO sectors.

Helen Marrison graduated from the University of York with a degree in Politics and a Masters in Cultural Heritage Management. She worked in the Social and Economic Research Team at Historic England (formerly English Heritage), responsible for analysing trends in heritage engagement in England in order understand the impact of heritage on communities, individuals and the economy. She was involved in the production of the 2014 and 2015 Heritage Counts reports, and coordinating the Heritage Champions programme. Helen now works for a leading property consultancy in London as a planning and heritage consultant.

Aya Miyazaki is a PhD Candidate in International Relations at the University of Tokyo and a senior officer at the Japan Foundation. A graduate of the Conservation Studies (Historic Buildings) MA programme, University of York, her research focuses on the system of conservation and sustainable management of cultural heritage. She is a member of Young Heritage Experts, established by the 32 delegates of the Young Experts Forum 2015 of the World Heritage Committee, and ICOMOS Japan, in which she works as a member of the working group for the 50th Annual General Assembly of ICOMOS.

Nerupama Y. Modwel leads the Intangible Cultural Heritage Division at the Indian National Trust for Art and Cultural Heritage (INTACH), New Delhi, India. She has overseen the successful completion of a number of research, documentation and workshop projects on performing traditions in dance, drama and puppetry, folklore and oral narratives, and indigenous arts and crafts. She is also coordinating an ongoing programme of studies on tribal communities across the country, and cultural resource mapping of cities like Varanasi and Gaya. Her publications include a manual on intangible heritage collection, and books on folklore and mediaeval cuisine.

Sophie Norton is a building conservation professional with an active interest in education and training through conservation. As Heritage Skills Coordinator in York, she developed live-site conservation projects, work-based learning placements, a well-received programme of CPD courses and contributed to teaching on undergraduate and postgraduate programmes. Sophie's PhD research in conservation studies looks at the way the developing conservation profession has influenced communities of craftspeople since the beginning of the twentieth century. She currently works as a Training Officer with Historic England.

Krupa Rajangam has an MA in Heritage Conservation from the University of York (2005) and BA in Architecture from RV School of Architecture,

Bengaluru. Her practice, Saythu: Linking People and Heritage, bridges building conservation practice and community awareness of heritage. Krupa is Visiting Faculty at DSCE, MSRIT and BMS Schools of Architecture in Bengaluru for undergraduate conservation elective and design studios. She is currently pursuing her PhD on Social or Community Value in the Conservation Process through a Fellowship grant at the National Institute of Advanced Studies (NIAS), Bengaluru.

Henry Russell is Programme leader for the MSc in Conservation of the Historic Environment at the School of Real Estate and Planning, University of Reading. He is also course director for three short programmes at the University of Oxford. He is a chartered surveyor, specialising in the historic environment. After spending the first part of his career in practice, he has been engaged in teaching and research for the last 15 years.

Alaina Schmisseur holds an MSc in Conservation from the Institute of Archaeology, University College London. She has worked in the field and behind the bench in conservation and education for a variety of employers, including the York Archaeological Trust, SWCA Environmental Associates, the National Museum Wales, the National Museum of Science and Industry Group and AOC Archaeology. Since 2013 she has been employed by the Society for the Protection of Ancient Buildings as the North East England Project Officer for the Maintenance Co-operatives Project. Alaina is currently undertaking a PhD in Archaeological Conservation at the University of York.

Laura-Melpomeni Tapini studied Conservation of Antiquities, specialised in Mosaics, Stones and Frescoes, at the Scuola per il Restauro del Mosaico in Ravenna, Italy; later receiving her MA in Conservation Studies (Historic Buildings) from the University of York, UK. She worked for many years at different sites in Italy, Greece and Syria. She is the co-founder and, currently, Managing Director of DIADRASIS, organising educational, project-based workshops and research programmes for heritage and culture.

Jane Thompson is a member of the senior advisory body in the Italian Ministry of Culture and Tourism and one of the co-authors of the 2015 Sustainable Development policy for World Heritage and the 2013 UNESCO resource manual *Managing Cultural World Heritage* in her role as consultant to ICCROM. She has been Project Manager of the Herculaneum Conservation Project since this public–private partnership was launched in 2001 and is Course Director for Heritage Management at Bocconi University. She also consults internationally, providing heritage management expertise and capacity building for World Heritage properties in Europe and beyond.

Nigel Walter is a Specialist Conservation Architect and director of Archangel, a Cambridge- based practice specialising in churches and historic buildings. His first degree in Architecture at Cambridge University was followed by study for an MA in Conservation Studies (Historic Buildings) from the University of

York, where he is now undertaking PhD research ('All is Not Loss': Change, Narrative and the Community Ownership of Historic Buildings). His particular interest is in enabling church communities to engage better with their living buildings, and his most recent publication is the *Buildings for Mission* handbook (Canterbury Press 2015).

Gamini Wijesuriya qualified in Architecture and worked as the Director of Architectural Conservation of Sri Lanka (1982–99). He broadened his scope with an MA in History/Historic Preservation from Carnegie-Melon University (USA), an MA in Cultural Heritage Management from York University (UK) and, finally, his PhD from Leiden University (Netherlands). Since 2004, Dr Wijesuriya has been a Project Manager in ICCROM's Sites Unit, where he coordinates the course on the Conservation of Built Heritage, is deputy coordinator for World Heritage activities and oversees the Promoting People-Centred Approaches to Conservation programme.

Craig Wilson has 20 years of academic experience as a researcher in post-war reconstruction of cultural heritage at the University of York and as a senior lecturer in heritage and conservation at Northumbria University in Newcastle. He has represented the Twentieth Century Society on the Newcastle Conservation Advisory Panel and was active on the Byker Listing Working Group. He is a member of the council of management of the Tyne and Wear Building Preservation Trust, a trustee of the Britannia Panopticon Music Hall, Glasgow, and sits on the Society for the Protection of Ancient Buildings (SPAB) Scotland committee. He currently lectures freelance.

Preface

The collected papers in this volume are drawn from a conference held in York in 2014 and developed by their authors as contributions for this volume. The conference – Engaging Conservation: Communities and Capacity Building in Conservation Practice – was part of a wider project to examine critically how conservation practitioners are adapting to reflect a growing body of experience in wider heritage work and more people-centred practice in conservation. Other elements of the wider Engaging Conservation project included workshops on capacity building, on use of digital heritage, a day school on archaeological stewardship (Adopting Archaeology – now a collaborative project with the Council for British Archaeology) and a field project around the Church Conservation Trust's project at Holy Trinity Goodramgate in York.

A wide range of heritage organisations took part: local preservation trusts and amenity groups, open air museums, national bodies – the Heritage Lottery Fund, English Heritage (now Historic England), Council for British Archaeology, SPAB – and included colleagues from Europe (ICCROM in Rome), India (INTACH) and York conservation alumni from around the world. Chapter 1, as an introduction, draws on the views and wide-ranging experience shared and exchanged as part of this project.

This volume presents both research and practice perspectives in community-led conservation. Many issues debated in the workshops and conference remain open ended, contested notions and definitions of 'community', of expert roles and of the nature of capacity building: others emerged as clear opportunities for performing conservation in new ways and are encouragingly already appearing in practice in new heritage projects. We have much to learn from our colleagues internationally.

Acknowledgements

I would like to acknowledge with gratitude a number of organisations and individuals who contributed to the success of the Engaging Conservation project which made this publication possible:

The University of York for financial resources through its External Engagement Fund;
The York Conservation Alumni Association and Peter Gouldsborough for their support;
Pam Ward for skilfully coordinating the conference administration and Gavin Ward for design work;
Keith Emerick and Danai Koutromanou for their supportive collaboration throughout.

Finally, my personal thanks go to all the contributing authors for their time, insightful writing and patient forbearance while we assembled the final papers.

Gill Chitty, Editor

1   Introduction

Engaging Conservation – practising heritage conservation in communities

Gill Chitty

Public participation and local community involvement in heritage conservation have taken centre-ground over the last two decades. Arguably recognised as the new norm, and as established principles in planning and cultural policy, they are less well embedded in contemporary conservation practice. In the UK, *Power of Place* (English Heritage 2000) was a watershed for this change in focus, although internationally the shift towards recognition of cultural pluralism came much earlier, underpinning the evolution of the Australia ICOMOS *Burra Charter* from 1979 and international principles on social inclusion and diversity embedded in conventions and charters from that decade onwards (ibid.; Australia ICOMOS 2013; UNESCO 2005). The current UK Government White Paper on Culture, just published at the time of writing, acknowledges still 'the challenge of creating more cultural opportunities, particularly for those whose chance to experience culture is more limited' (DCMS 2016:13). However, its focus is firmly on what 'culture' can do for society and the economy rather than on recognising and nurturing what communities do to enrich and creatively shape their own culture and heritage. In this volume, drawn from the 2014 Engaging Conservation conference and collaborations at the University of York, we bring together ideas and experience from people working in cultural heritage conservation to look critically and honestly at the progress we have made – see, for example, Keith Emerick on 'Ending the tyranny of Ruskin and Morris?' (2014: 219–37) – in democratising our practice and the changes that are still to be negotiated.

Unquestionably, with a shift in emphasis towards privileging social values and inclusive heritage processes, conservation practitioners are increasingly aware of this agenda. The movement towards it has been a gradual one. At the international level, it might be traced back to Local Agenda 21 policies, enacted in the years following the Rio Earth Summit in 1992, with its focus on citizenship and communities acting at the local level in socially and environmentally sustainable ways. Working in government during the 1990s, no one could be unaware of its direct impact on both central and local policy in the UK, reflected by environmental and heritage government agencies (Countryside Commission *et al.* 1996; English Heritage 1997). This direction of travel was encouraged by debate on the creation of public value: Mark H. Moore's work at Harvard, its appropriation in

New Labour public policy and a re-examination of the value of culture to society in the mid-2000s (Moore 1995; Jowell 2004; Holden 2004; Clark 2006). At international level, the ICOMOS *Stockholm Declaration* (1998), recognising 'the right of everyone to partake freely in the cultural life of the community', was soon followed by the *Council of Europe Framework Convention on the Value of Heritage for Society, Faro* (Council of Europe 2005) and the UNESCO *Convention on the Protection and Promotion of Diversity of Cultural Expressions* (UNESCO 2005). The global movement in thinking about culture, society and sustainability is outlined more fully below, in Chapter 3, 'People-centred approaches' (Wijesuriya, Thompson and Court, pp. 34–49), and is reflected in new *Principles for Capacity Building* for ICOMOS International Training Committee (CIF 2013). Community-driven conservation and local empowerment have become major themes for the international community 'placing people at the centre of the cultural debate' (ICOMOS 2014: 2).

Locally led, active participation and social relevance are, then, dominant characteristics of the twenty-first century's democratised cultural heritage practice. Whether at global or at local scales, it is recognised as being, above all, about human interactions. Evaluation of its outcomes in the UK shows demonstrable social, educational, well-being and personal benefits for participants, as well as tangible outcomes, which are being carefully measured and critically examined (e.g. HLF 2015; Neal 2015). This sits alongside an expanding body of critical and theoretical literature around authority and heritage work, especially in archaeological practice (Graham *et al.* 2000; Merriman 2004; Smith 2006; Smith and Waterton 2009; Waterton and Watson 2011; Moshenka and Dhanjal 2012) and as considered by other contributors to this volume. The role of the expert practitioner in heritage work continues to be problematised, questioned and repositioned – see for example see John Schofield's *Who Needs Experts: Counter-mapping Cultural Heritage* (Schofield 2014).

Community engagement in conservation practice, in contrast with the community focus in archaeological and heritage work, has received relatively little attention in terms of widening engagement until recently. By 'conservation' here, the focus is on the care of tangible cultural heritage: buildings, structures and remains in which material interventions are made with the intention of safeguarding and revealing aspects that are valued. The most recent iteration of the *Burra Charter* retains its classic definition of conservation as 'all the processes of looking after a place so as to retain its cultural significance' (Australia ICOMOS 2013: 4). But, just as heritage is contingent and dynamic, perpetually newly created in practice, so it is in conservation. In an initial project workshop on Engaging Conservation, participants arrived at a definition of conservation 'as the management of creative continuity and socially cohesive heritage practice (rather than the management of change)'. In discussion following his keynote (below p. 31), Jukka Jokilehto suggested that there might be 'less emphasis on managing change, more on managing continuity through change', and a move for conservation practice towards what Smith has characterised in heritage work as a form of 'social practice':

The product or the consequences of heritage activities are the emotions, experiences and memories of them that they create... What is also created, and
indeed continually recreated, are social networks and the historical and cultural narratives that underpin these binding relations.

Smith 2012: 3

Conservation work might, for example, be reframed with its community networks
as actors, co-creating material forms and new participative experiences of (re)
making heritage through its conservation, as explored in the ethnographic work
of Jones and Yarrow (2013: 23). In established practice, conservation has been
mainly concerned very practically with methods and treatments for structure,
monument or place, and the outcomes of that intervention. But, as Erica Avrami
has questioned: 'what if conservation principles focused instead on how people
engage in decision-making about their heritage and the *process* of participation?'
(Avrami 2011: 182, emphasis original).

Notwithstanding some excellent, innovative working in this area, conservation
practice in the built historic environment still remains largely an expert, professional domain: a field in which specialist practitioners and decision-makers
consult with local people and (sometimes) facilitate their involvement. A reading
of the present UK Heritage Lottery Fund website, as a point in time for this
critique, illustrates the separation that heritage conservation practice still perpetu-
ates. 'Community heritage' is about people: 'projects that help local people delve
into the heritage of their community, bring people together, and increase their
pride in the local area'. It includes actions like 'revive a long-lost tradition or
craft... share forgotten folk-tales... celebrate the lives of people commemorated
on a war memorial... reveal the stories of the area's diverse groups... discover
the origins of the place where you live' (HLF 2016a).

'Conserving historic buildings' in contrast 'helps to save the places people
value and provides communities with much-needed facilities'. It engages interestingly both with the risk of loss and the value of utility, with material structures and practical tasks: 'repair and transform a historic building... help
volunteers try their hand at building conservation... take a historic building off
a Buildings at Risk register... regenerate a historic town centre' (HLF 2016b).
These are tasks that might well seem beyond the means or capacity of a local
society or community group without bringing in the experience and knowledge
of others.

The focus of the Engaging Conservation workshops and conference was, therefore, around the question of growing more meaningful community engagement
in this area. How can we be more successful in building capacity for local ownership and leadership of heritage conservation projects, and genuine participation
in decisions and in practice? How effective are attempts at wider public engagement with what heritage conservation aims to achieve for public benefit? The
case studies explored in this volume are not framed as a 'models' for forms
of community participation but, rather like the case of Rione Sanità in Naples
(p. 43), as 'real-life examples [that] might help identify some key capacities

required for successful management'. Looking beyond the UK to examples in Europe and to Asia provides innovative and fresh perspectives on conservation as cultural practice.

In the years ahead, we can expect there will be increasing reliance on the voluntary and community sector. Given the prevailing economic environment, public resources for heritage conservation have been and will continue to be significantly reduced, nationally and globally. Through the Engaging Conservation project we wanted to look at the challenges and benefits in this new landscape of local activism. What works well and why? What we can we learn by sharing approaches and practical experience at a local and at international level? In workshops we explored notions and definitions of community that are problematic. They can be self-defined, exclusive and potentially self-interested or parochial. Equally, they may be empowering, locally legitimate and altruistic in their goals. However, it was also clear that expertise is valued. It may not be that the role of the conservation expert is over, but certainly time that expert roles should be renegotiated: 'as commodity rather than authority, as facilitator rather than fixer?' (Chitty 2014).

Teaching heritage conservation to practitioners and postgraduates it is evident that we lack the same body of literature and critical theory around material conservation practice as has developed around heritage practice (as Nigel Walter explores, below p. 51). Looking for a recent critical overview for student reading, and finding very little, was an impetus to gather this thinking and these conference papers for a volume marking a point in evolving practice. Moreover, understanding how practice must and is adapting to a climate of consensual, collaborative, citizen-centred endeavour, should be rooted, I would argue, in lived experiences. It is not enough to theorise or critique outcomes and attitudes, but to examine meaningfully what happens in or results out of practice, pushing boundaries, observing and learning.

This collection of papers from the conference has a strong York focus in the case studies and in the contributors, many of whom are based in York or practise here. The Engaging Conservation project uses the historic city as a site for creating new understanding of how we can build capacity. Evolving new ways for anyone who wishes to be involved in conservation to have a voice is necessarily grounded in local practice and networks, as compellingly retold in 'Living with history in York: increasing participation from where you are' (Brigham *et al.* below p. 143).

Approaches to community engagement, participation and capacity building

In Part 1 of this volume are nine chapters examining larger issues of principle and approach on these themes of community engagement and capacity building. Presenting different institutional and stakeholder perspectives, they range from international policy to critiques of current practice by a heritage architect and heritage professional. Approaches developed in India for documenting intangible cultural heritage and national policy in England, for example, offer contrasting positions, reflective of distinctive cultural traditions and institutional practices.

Jukka Jokilehto's opening 'think piece' is written from a lifetime's experience in international heritage conservation working with the World Heritage Committee. His historical perspective on the evolution of community-led and conservation practice in a World Heritage context reflects on the questions around engaging conservation with societal values: that 'aesthetic recognition of a work of creativity in human consciousness, the significance of an element of heritage within its cultural context, and the diversity of cultural expressions — are all fundamental to what is at stake when discussing the role of heritage community in the management of its place' (Chapter 2, p. 22). Using case studies from Azerbaijan to Yemen, he illustrates how effective conservation and community involvement have (or indeed have not) interacted on the international stage. Conservation of heritage is not feasible without this interaction but, as he highlights, now demands recognition of the interests of multiple communities of stakeholders, a careful understanding of the complex interplay between them and the place of capacity building as an instrument of social equity. How should the role of the expert change in such initiatives and is the evolution of skilled expert facilitators to work in this new landscape of conservation practice keeping pace?

This question is the central concern in the keynote paper from Gamini Wijesuriya, Jane Thompson and Sarah Court (Chapter 3), illustrating the important role that ICCROM[1] has played, and continues to develop, globally in this arena. Engaging communities, they write, is about strengthening their ability to participate meaningfully in the process of making conservation and management decisions for themselves and their heritage, *and* in the process of implementing them. ICCROM, as one of the three advisory bodies to the World Heritage Committee, has pioneered new approaches and its Living Heritage, people-centred principles place the living dimension at the core of decision-making. Looking at capacity building in terms of the World Heritage management system – and this is applicable to heritage management at any scale – the approaches presented here encourage heritage practitioners to frame every specific conservation process as a contribution to society, in social but also in economic and environmental terms. The case study of Rione Sanità in Naples illustrates how local community action is transforming a neighbourhood with benefits for both the residents and the heritage. Communities can no longer be considered as just another 'category of stakeholders, but must be a *sine qua non* within the heritage discourse' (see p. 38) in which practitioners arguably have a subsidiary role as facilitators and technical advisers.

Nigel Walter's engagingly titled 'Everyone loves a good story' (Chapter 4) investigates further the paradox implicit in the role of professional practitioner and community enabler: that conservation practice, in remaining expert-led, is increasingly out of place in a climate of greater public participation. Conservation, he argues, with its focus on managing change, lacks an equally rigorous rationale to take account of how heritage is created, and therefore of how change can positively enhance an evolving and inclusively owned historic environment. Drawing on his professional experience with parish communities in England, Walter takes the work of Poulios (2014) with the communities of a group of monastic sites at Meteora, Greece. Poulios envisages a shift to viewing communities and sites as an inseparable entity, moving the focus of conservation from 'preservation' towards a continual process of 'creation'. Walter sets the orthodoxies of expert professional practice alongside community engagement with the adoption of plural, value-based conservation approaches, and highlights the tensions between the dual epistemologies of heritage studies and conservation practice. Arguing that the communal value of an historic building is better understood as an ongoing community narrative, rooted in dynamic living traditions, he proposes an alternative, narrative approach which makes space for ongoing cultural (co) production in a community, together with its expert practitioners.

Keith Emerick foregrounds the growing consensus on need for a radical shift in the way that conservation practitioners act and work with volunteer communities in a different UK context, and critically examines whether the public rhetoric of participation and community engagement has given rise to any genuine change in practice. In Chapter 5, he identifies the same fundamental disjunction between the epistemologies of building conservation and of heritage as social practice which have yet to be resolved, and reflects on whether value-based approaches to conservation, adopted internationally since the early 2000s (Clark 2014), are being realised effectively, or indeed at all, in practice. The experience of two community conservation projects in North Yorkshire highlights scope for the creation of new imaginative and 'trustful' relationships in locally led participation.

In Chapter 6, 'Community involvement in cultural mapping and safeguarding of intangible cultural heritage', Nerupama Y. Modwel shares an innovative Indian approach to involving local communities in documenting a rich intangible heritage of religious practice, craft, music, dance and creativity. The Indian National Trust for Art and Cultural Heritage (INTACH) is pioneering initiatives to define and map intangible heritage practices with communities, involving local volunteer facilitators and project leaders from its regional INTACH Chapters. The interaction between the documentation process and the community's intangible heritage is negotiated with community members: INTACH facilitators participate not as experts but as collaborators. By providing local workshops in individual villages and neighbourhoods, members of the community itself form part of the core research and survey team and enact self-documentation of the heritage practices they identify themselves. Their involvement in the process creates its own energy and the carefully structured, lengthy process – workshops, documentation and support for continuing community-based practice – is designed to create a sustainable social and economic future for both the community and its traditions. As a national organisation, INTACH is able to use its network of regional Chapters to find ways to operate effectively at a local level, even in a major subcontinent, using and adapting a common methodology with the communities themselves.

Returning to the UK context in Chapter 7, insights into Historic England's perspective on community-led conservation reflect both the constraints and potential future for a national heritage agency for engaging with community and local agendas. Helen Marrison's account of current initiatives examines the scope of national research (the work of Historic England's social and economic policy unit complements the very significant research undertaken for the Heritage Lottery Fund (HLF 2015)), its support for locally led projects and participation, and possible directions for future policy. Facilitation of community projects from a national level is necessarily at arm's length: resourcing pilot studies, sector research and best-practice guidance. Historic England's heritage-at-risk pilot surveys, reporting on the condition of Grade 2 listed buildings, provide an interesting case study, carried out by local groups and societies using centrally distributed documentation and a digital platform. The initiative is now being developed for a nation-wide volunteer programme in partnership with Civic Voice. It offers, perhaps, one model for future transition from centralised to locally enacted practice. Could national monitoring and reporting on the condition of heritage assets move towards wholly devolved practice in which agency in the heritage sector passes into the hands of voluntary local groups? Historic England was established with a new identity in 2015 and the new national body's role in community engagement and conservation comes at an interesting moment in defining its fresh direction.

The effectiveness of national initiatives as catalysts is exemplified in Ian Bapty's Chapter 8, on the work of the Industrial Heritage Support Officer. This Historic England-supported project addresses the sustainability of the voluntary sector communities who are prime movers in conservation and preservation of industrial heritage. As Bapty notes, this is a particularly insightful context in the current climate, when the care of heritage is generally moving towards 'people-led' management models. The history of preservation of industrial sites in Britain is arguably an exemplar of 'bottom-up' community-based, heritage management: it also displays many of the practical difficulties and limitations that are still experienced in community conservation practice. This project explores practical and new ways to reinvigorate and sustain an existing volunteer-based tradition of looking after industrial sites. In a locally led and managed future for heritage assets, understanding how a mature community of local volunteer expertise can be reinvigorated and bring in a fresh generation of young activists will be valuable learning. Bapty provides an impressive historic review of the industrial heritage preservation movement. It raises an interesting aspect of early twentieth-century concerns with securing the living heritage of working people's traditions and experience, as much as the material culture of industry. Now on the cusp of a generational change, the legacy of living heritage (i.e. where the traditions and knowledge of working industrial practice are within the lifetimes, if not the lived experience, of volunteer groups) is in transition to a generation separated from it, bringing into sharp the relief the need to manage the heritage of craft and trade practice itself. How do conservation practice and traditional craft practice interact in this field? For example, could some of the apparent misunderstanding about 'the differences between conservation and restoration of historic machinery' (p. 112) attach to deeper differences between a heritage of traditional practice and the practice of heritage conservation? The issues for the future, around succession planning and training, reflect the same economic factors that appear in later case study chapters (such as the Castleford Heritage Trust in Chapter 11), the disadvantaged status of smaller charities and the partnership collaborations that are central for sustaining small independent, volunteer enterprises.

The social and economic benefit of cultural engagement and participation remains a central concern for funders and politicians (DCMS 2016: 15), and the question of how the outcomes of such engagement might be measured or evaluated is examined critically in Chapter 9, in Danai Koutromanou's case study of visitor experience of conservation in a museum context. Her findings highlight the need for conservation practitioners to consider public views about heritage conservation, both as a social process of engagement and as an important aspect of professional practice. The impulse to use evaluations to demonstrate the success of projects – driven by the undeniably important goals of satisfying funding bodies and institutional objectives – is having the effect, she argues, of limiting the value of research into public experience of heritage conservation. More nuanced and exploratory use of mixed methods, as illustrated here, could aid better understanding of how to communicate more meaningfully with, and respond to, the public audiences for whom conservation is performed. Rather than thinking about the evaluation *of* success, too often expressed in terms of ill-defined outcomes, this research suggests reorientation to focus on how to design evaluation *for* success in dynamic ways that could actively inform practice.

Case studies: Engaging Conservation in community practice

The case studies in Part 2 of this volume contextualise current debates in community conservation and examine diverse contexts across England, Spain, Japan and India. All nine chapters are about recent and ongoing projects where the authors are actively involved, and play a range of roles: as participants, professional advisers and facilitators, community leaders and coordinators. Some chapters present commentary and critical review; others voice the lived experience of conservation practised in and with communities, with immediacy and innovation. Our understanding of participation and activism in conservation practice is being changed with the use of ethnographic and action research methods and of approaches developed in other forms of heritage work and in other sectors. These individual studies bring insights into the implications of what an inclusive and community-led form of heritage conservation looks and feels like from the local level: changed perspectives, priorities, shared decision-making, risk taking, boundary-breaking, new networks and unexpected outcomes.

In Chapter 10, 'Living with history in York: increasing participation from where you are', the four co-authors – Lianne Brigham, Richard Brigham, Peter Brown and Helen Graham – write about the York strand of a UK-wide research project: 'How should heritage decisions be made?'. It explores, and exemplifies in itself, alternative ways of sharing views and heritage decision-making processes among local people in York. The chapter adopts a new form of telling, of co-creating an account of the project in which the authors 'hope we can go beyond telling you how we've worked together but instead *show* you – model it – through the way the chapter is put together' (p. 143, emphasis original). It is not always helpful to think about professionals and communities, they suggest. 'Instead, we can all grow networks of people across boundaries who can constantly challenge you to think differently and refuse to let you off the hook and think that nothing can be done' (p. 161): inspirational and innovative work.

Moving to a distinctly different historic Yorkshire town in Chapter 11, 'Community heritage and conservation in Castleford, West Yorkshire', Alison Drake presents a perspective from a community heritage trust. Faced with high levels of social deprivation in an economically disadvantaged, former mining community, the Trust is shaping its own heritage-led regeneration. This is a passionately honest, personal account of its aims and ethos, struggles and aspirations for the Queen's Mill, the world's largest stoneground flour mill, and the Trust's 'longstanding aim to reclaim and use our tangible and intangible heritage and culture to build a better life for our community' (p. 163). The mill is not just a heritage regeneration project. For the Trust and the Castleford community this is about more than simply providing a heritage centre for a market town: 'it is about trying to turn around the fortunes of a whole community, it is about building a better future for our people' (p. 173). It is also about having their own agency beyond the decision-making process, 'to conserve, use and own at least one tangible aspect of our varied, unique industrial and social heritage' (p. 163).

Alison Drake voices the immediacy of need, the setbacks and obstacles the Castleford community has negotiated and tells the stories of its skilful use of local and national political agendas and media to serve its purposes. Like York – yet so unlike it in many ways – we hear that it is through building networks of common interest, the 'fragile coalitions and alliances which need effective negotiators' (p. 176), that new solutions are being found. Anyone who works for a local authority or heritage funding body should read this to understand what running a small community conservation charity *feels* like. Conservation in practice here is about developing skills in groups and people management, generating honesty and trust in working together to create equity in new partnerships. It is also about the obstacles that face a community that strives for ownership of the whole conservation process, as well as the heritage asset itself.

An industrial heritage conservation thread runs on through the next two chapters and illustrates the powerful connections that it makes: reimagining and repurposing the built heritage of past working communities brings social meaning as well as practical and sustainable reuse for contemporary ones. These are case studies at opposite ends of the scale: the massive timber, riverside wharfage of the Dunston Staiths (in Chapter 12) and the domestic scale of back-to-back housing in Leeds (in Chapter 13). Craig Wilson tells the story of local empowerment that underpins the successful conservation and creative use of a major monument of the north-east coal industry, Dunston Staiths. Starting out as a classic top-down government initiative in the Garden Festival of 1990, the future for this iconic structure of the Tyneside coal industry has been secured, ultimately, through its reinvention as a community asset and platform (literally) for cultural actors. The synergies with the spirit of community and aspiration of the Castleford Mill project are strong here: a post-industrial community, high levels of deprivation, still scarred by the forced closure of the coal mining industry. Here in a larger scale, urban riverside setting – and working with a listed and scheduled heritage asset (how much does difference does and should this make?) – this project has also been driven forward through the strong leadership, vision and a wide-ranging partnership of local agencies and, strikingly, major cultural investment – a story of vibrant success and exciting possibilities. These projects foreground the rein-vention of conservation, as envisioned in the 2014 workshop, as 'management of creative continuity and socially cohesive heritage practice'.

Thinking about the places where people live comes to the heart of a community and Joanne Harrison's chapter examines recent regeneration programmes for terraced house communities in the north of England. She characterises this as the 'forgotten' agenda of terraced housing: its heritage value and significance for local people. Community engagement is well-established practice in developing housing and neighbourhood regeneration programmes, but heritage and conser-vation issues very rarely appear in the frame in developing proposals with communities who *live* in Victorian terraced housing. Harrison's chapter critically reviews how the ideals for participatory engagement by communities in housing renewal have been enacted in practice and how little heritage values have figured, if at all, in this process. She charts the campaigns of the Welsh Streets Home Group in Liverpool to halt the Pathfinder housing clearance programme and introduce a sustainable programme for renewal of traditional terraced houses, activism which finally triumphed in the award of last year's Turner Prize to the Granby 4 Streets CLT (2015). In the Leeds case study, effective engagement with stakeholders in the regeneration of their back-to-back housing was achieved but the voices of local working communities appear simply not to carry weight, even where they are able to be heard. Harrison's chapter concludes with options for improving the way in which renewal of historic urban housing could take place with the communities who live in and value them. Clearly there are professional and policy agendas to develop here for effective conservation practice and work on a much-needed body of further research.

With an international gaze in the next two chapters, it is insightful to see the same issues around voicing of community values being explored in India, in a very local, neighbourhood context, and in Japan, in relation to the communities of World Heritage sites. Krupa Rajangam's practice Saythu has developed initia-tives in community-centred projects in Bangalore engaging local perspectives in mapping heritage – *Nakshay* and as *Neighbourhood Diaries* – discussed in Chapter 14. This is explorative, reflexive work – asking questions, trialling meth-odologies to examine whether established models and roles in professional prac-tice are relevant – finding the unexpected in communities that may be tentative in judging motives, outcomes and framing boundaries, less certain about the value of their heritage. As other community research in India is showing, there can be no standard approach to ensure meaningful enactment of community heritage

(Sharma 2013: 282); but the thought-provoking evaluation here points the way towards the importance of personal associations and experiences, of multi-disciplinary practice and of careful attention to local voices.

Aya Miyazaki writes of her research with remote rural communities in Japan and the interrelationship of local, national and international values in Chapter 15. In these case studies of Shirakawago and Gokayama villages and Iwami Ginzan Silver Mine cultural landscape, personal and community activism in decisions about village matters at a very local scale interacts with the management of globally recognised heritage. The role of municipal and national government here is instrumental in effective engagement, but it is the communities themselves who have taken charge of protecting their cultural heritage sites, creating their own set of conservation rules, as well as places where their voices can be heard and reflected in municipal plans for conservation management.

In Chapter 16 Laura-Melpomeni Tapini and Lucía Gómez-Robles reflect on their collaborative project in a small Spanish town, Martos, engaging local people, the local crafts school and the municipality in the conservation of a sixteenth-century fountain. Here they present their detailed methodology, in an international workshop, for fostering integrated involvement of local townspeople, not only in conservation practice, but also in actively deciding the future for the fountain as a community asset and helping to create sustainable prospects for it. Turning public curiosity into community interest, engaging local skills and leaving a legacy, the project demonstrates that meaningful practice in this field is carefully planned and enacted to be effective.

Turning back to the UK, in Chapter 17, on Maintenance Co-operatives, Stella Jackson and Alaina Schmisseur write about the Society for the Protection of Ancient Building's current capacity building programme for church parish communities. With the examination of methodology, here too is reflective comment on the nature of participation and the importance of creating a space where all kinds of participants can find a role to fit their circumstances. Key to meaningful participation should be development of new volunteer pathways and the (?self-)creation of roles bespoke to the ways in which people can, and want to, take part in heritage conservation practice, as complex, diverse and changeful as people are themselves. Viewing the same issues for parish communities in a broader perspective, in Chapter 18, Henry Russell and Philip Leverton consider capacity building in the professional sector and the need for new skills for all participants, as conservation practitioners, facilitators, enablers and informed clients.

Finally, returning to a close focus on conservation practice, Sophie Norton's Chapter 19 observes the experience of a community of apprentice stone masons working on a training project to conserve a traditionally constructed vernacular building, the 'Stone Masons' Lodge'. Here she highlights the diversity of values within communities of conservation practice themselves, reflecting on the insights that recent ethnographic research has brought to understandings of the interplay between them. Careful attention is paid to consultation and local involvement in decisions about the use and future care of heritage, but, during the actual process of conservation itself, as detailed decisions are made on site, another complex set of values and choices play out among different communities of interests.

This returns to thinking about living heritage practices being so fruitfully explored in ICCROM's people-centred approaches, which recognise that 'communities have capacities and assets that outlast political or professional structures and complement specialist knowledge and skills in managing heritage' (p. 36). It is in this context that the custodianship of communities in heritage conservation (with their traditional and local knowledge systems) can be seen as inherently sustainable. Conservation practice *is* moving, from a focus on the preservation of heritage and cultural material to encompass the value of the process itself, for society and for affirming individual and group identity in a place: it will need and bring new capacities, social networks and, challengingly, co-creation of new heritage in that process, as the chapters that follow will explore.

Note

1   The International Centre for the Study of the Preservation and Restoration of Cultural Property.

References

Australia ICOMOS 2013. *The Burra Charter: The Australia ICOMOS Charter for Places of Cultural Significance* [online]. Available at http://australia.icomos.org/wp-content/uploads/The-Burra-Charter-2013-Adopted-31.10.2013.pdf [accessed 28 August 2015]

Avrami, E. 2011. 'Heritage, values, and sustainability'. In A. Richmond and A. Bracker (eds), *Conservation: Principles, Dilemmas and Uncomfortable Truths*. London: Elsevier, in association with the Victoria and Albert Museum, 177–83

Chitty, G. 2014. Reflections on the introductory workshop, Engaging Conservation blog [online]. Available at https://engagingconservationyork.wordpress.com/2014/03/10/reflections-on-the-introductory-workshop/ [accessed 28 August 2015]

CIF 2013. *Principles for Capacity Building through Education and Training in Safeguarding and Integrated Conservation of Cultural Heritage*, Paris: ICOMOS CIF [online]. Available at http://cif.icomos.org/pdf_docs/CIF%20Meetings/Guidelines/ICOMOS_CIF_PrinciplesCapacity_EN_20130930.pdf [accessed 28 August 2015]

Clark, K. 2006. *Capturing the Public Benefit of Heritage*. Swindon: English Heritage

Clark, K. 2014. Values-based heritage management and the Heritage Lottery Fund in the UK. *APT Bulletin*, 45(2/3), 65–71

Council of Europe 2005. *Council of Europe Framework Convention on the Value of Heritage for Society, Faro, 27.X.2005*. Strasbourg: Council of Europe Treaty Series 199 [online]. Available at http://www.coe.int/en/web/conventions/full-list/-/conventions/treaty/199 [accessed 28 August 2015]

Countryside Commission, English Heritage and English Nature 1996. *Ideas into Action for Local Agenda 21*. Hampshire: Countryside Commission

DCMS 2016. *The Culture White Paper*. London: Department for Culture Media and Sport [online]. Available at https://www.gov.uk/government/publications/culture-white-paper [accessed 24 March 2016]

Emerick, K. 2014. *Conserving and Managing Ancient Monuments: Heritage Democracy and Inclusion.* Newcastle: Boydell

English Heritage 1997. *Sustaining the Historic Environment.* London: English Heritage

English Heritage 2000. *Power of Place.* London: English Heritage

Graham, B, Ashworth, G.J., and Tunbridge, J.E. 2000. *A Geography of Heritage: Power, Culture and Economy.* Oxford: Oxford University Press

Granby 4 Streets CLT 2015. History of the four streets [online]. Available at http://www. granby4streetsclt.co.uk/history-of-the-four-streets/ [accessed 24 March 2016]

HLF 2015. *The Values and Benefits of Heritage: A Research Review.* Heritage Lottery Fund [online]. Available at https://www.hlf.org.uk/values-and-benefits-heritage [accessed 28 August 2015]

HLF 2016a. What we fund/community heritage [online]. Available at https://www.hlf.org. uk/looking-funding/what-we-fund/community-heritage [accessed 28 August 2015]

HLF 2016b. What we fund/buildings and monuments [online]. Available at https://www. hlf.org.uk/looking-funding/what-we-fund/buildings-and-monuments [accessed 28 August 2015]

Holden, J. 2004. *Capturing Cultural Value: How Culture Has Become a Tool of Government Policy.* London: DEMOS

ICOMOS 1998. *The Stockholm Declaration of ICOMOS Marking the 50th Anniversary of the Universal Declaration of Human Rights* [online]. Available at http://www.icomos. org/charters/Stockholm-e.pdf [accessed 28 August 2015]

ICOMOS 2014. *The Florence Declaration on Heritage and Landscapes as Human Values Regarding the Values of Cultural Heritage in Building a Peaceful and Democratic Society.* ICOMOS General Assembly 2014, Florence

Jones, S. and Yarrow, T. 2013. Crafting authenticity: an ethnography of conservation practice, *Journal of Material Culture* 18(1), 3–26

Jowell, T. 2004. *Government and the Value of Culture.* London: Department for Culture, Media and Sport

Merriman, N. 2004. 'Introduction: diversity and dissonance in public archaeology'. In N. Merriman (ed.), *Public Archaeology.* London: Routledge, 1–17

Moore, M.H. 1995 *Creating Public Value: Strategic Management in Government.* Cambridge, MA: Harvard

Moshenka, G. and Dhanjal, S. 2012. *Community Archaeology: Themes, Methods and Practices.* Oxford: Oxbow

Neal, C. 2015. 'Heritage and participation'. In E. Waterton and S. Watson (eds), *The Palgrave Handbook of Contemporary Heritage Research.* London: Palgrave Macmillan, 347–65

Perkin, C. 2011. 'Beyond the rhetoric: negotiating the politics and realising the potential of community-driven heritage engagement'. In E Waterton and S. Watson (eds), *Heritage and Community Engagement: Collaboration or Contestation?* London: Routledge, 115–30

Poulios, I. 2014. *The Past in the Present: A Living Heritage Approach – Meteora, Greece.* London: Ubiquity Press

Schofield, J. (ed.) 2014. *Who Needs Experts: Counter-mapping Cultural Heritage.* Farnham: Ashgate

Sharma, T. 2013. 'A community-based approach to heritage management from Ladakh, India'. In K. Silva. and N. Chapagain (eds), *Asian Heritage Management: Contexts, Concerns and Prospects*, New York: Routledge

Smith, L. 2006. *Uses of Heritage.* London: Routledge

Smith, L. 2012. 'Discourses of heritage: implications for archaeological community practice'. *Nuevo Mundo Mundos Nuevos/New World New Worlds* [online]. Available at http://nuevomundo.revues.org/64148 [accessed 6 July 2015]

Smith, L. and Waterton, E. 2009. *Heritage, Communities and Archaeology.* London: Duckworth

UNESCO 2005. Convention on the Protection and Promotion of Diversity of Cultural Expressions. Paris: UNESCO

Waterton, E. and Smith, L. 2011. 'The recognition and misrecognition of community heritage'. In E Waterton and S. Watson (eds), *Heritage and Community Engagement: Collaboration or Contestation?* London: Routledge, 12–23

Watson, S. and Waterton, E., 2011. 'Heritage and community engagement: finding a new agenda'. In E Waterton and S. Watson (eds), *Heritage and Community Engagement: Collaboration or Contestation?* London: Routledge, 1–11

Part 1

Approaches

Community engagement,
participation and capacity building

2 Engaging Conservation

Communities, place and capacity building

Jukka Jokilehto

Reflections on Engaging Conservation

The scope of this chapter is to reflect on the meaning of issues that are seen to be related to engaging conservation in society. This reflection necessarily makes reference to modernity as it has developed from the eighteenth century. There are many advantages with life in our contemporary world, but there are also some malaises. The Canadian philosopher Charles Taylor (1991) has identified these under three headings. The first is individualism: while in the traditional world the community was given collective purpose by inherited rules or traditions, the modern man has wanted to liberate himself from such rules, taking individual responsibility for his creative capacity. This has resulted in what Taylor calls disenchantment – that is, secularisation (see also Gauchet 1985). Disenchantment is connected to another malaise, the primacy of instrumental reason. Taylor writes: 'By "instrumental reason" I mean the kind of rationality we draw on when we calculate the most economical application of means to a given end. Maximum efficiency, the best cost–output ratio, is its measure of success' (1991: 5). This implies the political dimension of the malaises of modernity. The institutions and structures of industrial-technological society tend to severely restrict the choices of individuals, which can result in highly destructive actions. In order to address such malaises, there is need for an institutional reorientation, which also means a need for renewed attention to community.

The conservation of the past achievements of humanity is not necessarily a new concept. However, it has recently evolved into a particular discipline that is closely associated with the advancement of modernity. Even though there was early recognition that it is the responsibility of society as a whole to take care of human heritage, the disenchantment and instrumentality, noted by Taylor, seem to have dominated the decision-making process. Consequently, for example, the World Heritage Committee has given little or no attention to community in the justification of heritage nominations to the UNESCO List. It was only in 2012, on the occasion of the 40th anniversary celebrations in Kyoto, that the conference stressed the role of community:

> Only through strengthened relationships between people and heritage, based on respect for cultural and biological diversity as a whole, integrating both

tangible and intangible aspects and geared toward sustainable development, will the 'future we want' become attainable.[1]

It is in this same spirit that the General Assembly of ICOMOS, in Florence in November 2014, adopted the *Florence Declaration on Heritage and Landscape as Human Values: Declaration of the principles and recommendations on the value of cultural heritage and landscapes for promoting peaceful and democratic societies*:

> Community identity is rarely uniform or static but is a living concept that is constantly evolving thanks to an interplay of past and present in the context of current geo-political circumstances. Around the world, contrasting – and often conflicting – community identities are expressed through (and can be shaped negatively or positively by) the range of activities and service provision offered at cultural heritage tourist destinations that are intended to take advantage of the economic, social and cultural benefits of tourism.
>
> ICOMOS 2014

This Declaration should be seen in the context of a broadening in the recognition of heritage from 'monuments and sites' in the 1960s to cultural landscapes and non-protected heritage decades later.[2] Indeed, the adoption by UNESCO of conventions on the intangible cultural heritage in 2003 and the diversity of cultural expressions in 2005 have sustained the trend for an international recognition of the community's role in the processes of recognising and safeguarding heritage resources (UNESCO 2003, 2005).

Some working definitions would be helpful here. Community can be defined as a social unit that shares common value judgements. Community has been, indeed, a fundamental part of human existence ever since prehistory. It is within community that human creativity started finding its more permanent expressions of social unity and spiritual belief. In today's globalising world, the local community must take its responsibility for the recognition and care of its inheritance, which is also increasingly seen as a shared heritage involving a variety of stakeholders.[3]

Place is where the results of human creativity find their expression, carrying testimony to duration over time. Each place is characterised by its specificity resulting from the diversity of cultural expressions, as defined in the UNESCO 2005 Convention. Culture has a variety of complex meanings, which range from cultivation, such as agriculture, to maintenance, study and learning, worship and cult. Culture can be viewed as both the generator and a product of development within the evolving framework of the economy of a community. Economy is the system established by a community to provide the desired quality of life. It consists of labour, production, trade distribution and consumption, based on existing and/or newly generated resources. We can see the economy as a system within which a community arranges its resource management over time (Jokilehto 2012).

Heritage community

The concept of community can be seen as the basis for human existence. Traditionally, a community would have been related to a particular place, distinguished from its neighbours. It is here that each community developed its traditional specificity over time, associated with its rules of behaviour. It is also in this context that we need to understand the meaning and significance of inheritance – seen in its diverse dimensions, tangible and intangible, movable and immovable. Traditional society was also intimately associated with nature, which formed the context and which was also associated with meanings. Such issues are still preserved, for example, in the Okinawa islands of Japan, where local communities worship places of cult in nature, so-called *utaki* (Okinawa International Forum 2004). These could be small groves or woods, or even just a tree, associated with common values as sacred places. It is around such places that the community could find their common parameters and symbolic meanings. Indeed, heritage could be understood as a unifying feature for community and, for that reason, also subject to respect and care. At the same time, heritage was a carrier of knowledge embedded in the creativity and innovative solutions that became testimony to the learning processes by the community, not only internally but also in exchange with others.

In modern world society, particularly from the end of the twentieth century, there has been a certain dispersal associated with new systems of communication and new technologies, which have been associated with threats to the existence of traditional community. Indeed, the value judgements of today are less and less associated with traditions and increasingly with modern innovations offered by the globalising world. In some way, it seems that the traditional community is being increasingly replaced by 'virtual communities', which have their own canals of global communication, no more dozens of stakeholders but thousands and even millions of participants. In this context, the traditional heritage is gaining new significance.

In the history of heritage conservation, we can recall the Wonders of the Ancient World, referred to as remarkable constructions of antiquity by classical authors. They included the Great Pyramids of Giza, the only ones still standing today. In the Italian Renaissance, heritage was associated with important works of art or monuments representing history and knowledge of the ancient peoples. However, these were at risk of being destroyed and their material used in new construction. In response to this, in 1515, Raphael wrote a letter to Pope Leo X suggesting ways of halting the destruction of ancient monuments. He invited the Holy Father to:

> take care of what little remains of the ancient mother of Italy's glory and reputation; that is a testimony of those divine spirits whose memory still sometimes calls forth and awakens to virtues the spirits of our days; they should not be taken away and altogether destroyed by the malicious and ignorant who unfortunately have insinuated themselves with these injuries to those

hearts, who through their blood have given birth to much glory to the world and to this 'patria' and to us.

Cited in Bonelli 1978: 469

This letter resulted in a brief, signed by the Pope on 27 August 1515, appointing Raphael responsible for the protection of such monuments. It was one of the early preservation orders of the modern world.

Nearly five centuries later, the protection of heritage still requires being brought to the attention of the authorities, inviting them to sign consequent orders. Following a number of national legal instruments, especially from the nineteenth century, international concern for heritage was raised in the Constitution of UNESCO (adopted in 1945). Article 1 states that UNESCO is to:

> Maintain, increase and diffuse knowledge: By assuring the conservation and protection of the world's inheritance of books, works of art and monuments of history and science, and recommending to the nations concerned the necessary international conventions.

UNESCO 1945: Article 1.2c

In the European context, the *European Cultural Convention* of 1954 regarded the *languages, history and civilisation* of the different European peoples as their 'common heritage' (Council of Europe 1954). In 1972, the *Convention Concerning the Protection of the World Cultural and Natural Heritage*, however, noted that the cultural and natural heritage were increasingly threatened, not only by traditional causes of decay, but also by changing social and economic conditions which aggravated the situation (UNESCO 1972). Indeed, for various reasons, local communities have tended to forget or lose their traditions, replacing them with innovative modern solutions, thus challenging the traditional respect for heritage and ways of caring for it.

It is in this context that we should understand the *Framework Convention on the Value of Cultural Heritage for Society* (Council of Europe 2005). While previously heritage had often been labelled under different headings, such as movable or immovable, tangible or intangible, cultural or natural, the Faro Convention proposes a comprehensive definition:

> *cultural heritage* is a group of resources inherited from the past which people identify, independently of ownership, as a reflection and expression of their constantly evolving values, beliefs, knowledge and traditions. It includes all aspects of the environment resulting from the interaction between people and places through time.

Council of Europe 2005, emphasis added

This heritage is seen as an object of individual rights which give it meaning. Heritage is also seen as a 'source and resource' for the exercise of freedoms, putting 'people and human values at the centre of an enlarged and cross-disciplinary concept of cultural heritage'. The Faro Convention further continues to define the group of people who should have the principal responsibility of safeguarding such heritage. Thus, a '*heritage community* consists of people who value specific aspects of cultural heritage which they wish, within the framework of public action, to sustain and transmit to future generations' (ibid., emphasis added). These concepts can be taken as references for the development of appropriate means of protection and instruments for the management and valorisation of the common heritage, while respecting cultural and heritage diversities.

Creativity and community

If all human habitats were of the same type, with the same characteristics, and if they were recognised as heritage worthy of protection, their conservation could be relatively easily defined. Instead, humanity is characterised by its creativity, which not only defines *homo sapiens* as a human being with creative capacity, but also characterises the diversity of human cultures. The problem is that while recognising heritage as common for all humanity, the individual elements of this heritage are characterised by their creative diversity in reference to time and place. Therefore, while we propose common criteria for the management approach, the reality of a resource recognised as heritage is characterised by its diversity and specificity.

Undoubtedly, there will be common features in the approach to the conservation and restoration of objects or places recognised as heritage. Such a general methodological approach can be seen in art philosophy, as well as in the theory of restoration. The American philosopher John Dewey reflected that art has aesthetic standing only when it is experienced by a human being:

> A work of art, no matter how old or classic, is actually and not just potentially a work of art when it lives in some individualized experience. As a piece of parchment, of marble, of canvas, it remains (subject, however, to the ravages of time) self-identical throughout the ages. But as a work of art, it is recreated every time it is aesthetically experienced.
>
> Dewey 1934: 113

This means that a work of art has two components: the intangible, based on recognition, and the tangible that expresses the material aspect, which is subject to ageing. A work of art thus lives on and is recreated in the consciousness of the people who recognise it.

In the first decade of the twenty-first century, UNESCO has adopted two international conventions regarding cultural heritage. One was in 2003 for the *Safeguarding of the Intangible Cultural Heritage*, such as oral traditions, traditional crafts and so on. Different from the recognition of the outstanding universal value as indicated in the 1972 *Convention Concerning the Protection of the World Cultural and Natural Heritage*, the 'Operational Directives' of the 2003

Convention require that the *significance* of the nominated element be demonstrated – that is, 'Inscription of the element will contribute to ensuring visibility and awareness of the significance of the intangible cultural heritage and to encouraging dialogue, thus reflecting cultural diversity worldwide and testifying to human creativity' (UNESCO 2014: 28). At the same time, it is required that 'the element has been nominated following the widest possible participation of the community, group or, if applicable, individuals concerned and with their free, prior and informed consent' (ibid.: 27). The other convention was adopted in 2005 on the *Protection and Promotion of the Diversity of Cultural Expressions* (UNESCO 2005). This expands the previous definition of works of art to a broader concept: 'cultural expressions', defined as those expressions that result from the creativity of individuals, groups and societies, and that have cultural content. The 2005 *Convention* also refers to guiding principles, including particularly respect for human rights and fundamental freedoms, the principle of sovereignty, the principle of equal dignity of and respect for all cultures, as well as the principle of the complementarity of economic and cultural aspects of development.

These concepts — that is, the aesthetic recognition of a work of creativity in human consciousness, the significance of an element of heritage within its cultural context, and the diversity of cultural expressions — are all fundamental to what is at stake when discussing the role of heritage community in the management of its place. Thus, the community, in order to become 'heritage community', must be able to recognise the significance of the cultural expressions within the relevant traditional historical-cultural context, and its relevance to the present state.

In a case when the present-day community has kept the integrity and authenticity of the traditional continuity alive, the issue is mainly to provide the conditions for its continuity. Where such living traditions have been lost, which is only too frequent, and when the community does not have the necessary capacities, the building up of a heritage community becomes more complex. The two examples are very different. Nevertheless, in both cases, an important issue is the introduction of capacity building strategies with the integration of all stakeholders and the development and implementation of appropriate conservation and management programmes. The aim is to guarantee the continuity of the significance, in the first case, and the building up of consciousness of the significance, in the second.

Communities and place: an international perspective

Place is where the results of human creativity find their expression, as emphasised above, *carrying testimony to duration over time* – that is, the build-up of history that is contained in the layers of contributions by different generations. The management of places that have been recognised for their cultural significance depends on the specificity of each place. In principle, as is indicated in the 1976 UNESCO *Recommendation Concerning the Safeguarding and Contemporary Role of Historic Areas*, it is necessary to consider the totality of an historic area within its surroundings. Its specific nature depends on 'the fusion of the parts of which it is composed and which include human activities as much as the buildings, the spatial organization and the surroundings' (ibid.).

Each place has its own specificity, as has been indicated above. We cannot pretend that all management systems and plans would be exactly the same. Nevertheless, we can establish a methodology of approach which, in its general outline, is applicable to a variety of cases. Consequently, as indicated in the *Operational Guidelines for the Implementation of the World Heritage Convention*, 'an effective management system depends on the type, characteristics and needs of the nominated property and its cultural and natural context'. The common elements of management include a thorough understanding of the place by all stakeholders, the involvement of stakeholders and partners, capacity building, allocation of necessary resources, and monitoring and assessment of trends, changes and of proposed actions (UNESCO 2013: 27).

Even though the methodological approach may thus be based on the same ideas, starting from the identification and recognition of heritage significance, the cases vary. Unfortunately, in some cases, the traditional urban fabric has been destroyed due to business development and other ambitions, as in the case of the historic areas of Istanbul. These were inscribed on the UNESCO List in 1985, but now little remains of the original fabric (Figure 2.1). Also, the community has changed. The UNESCO website has described the situation as follows, and it is worsening:

> Vernacular timber housing in the Süleymaniye and Zeyrek quarters, was recognized as vulnerable at the time of inscription. Despite the threat of pressure for change, many efforts have been executed in order to conserve and strengthen the timber structures within the site since then. Changes in the social structure within the area have also affected the use of those structures. The urban fabric is threatened by lack of maintenance and pressure for change. The Metropolitan Municipality is attempting to rehabilitate the area to revive its degraded parts.
>
> UNESCO 1985

The Municipality has opted for a policy of replacement of the original traditional houses with new in a similar style, which however do not satisfy the requirement of authenticity. At the same time, the tendency is to rent such houses to wealthier families, and the earlier residents are gradually removed further outside from the historic centre area.

There are other more fortunate cases, where the community has been able to maintain their traditional habitat, as in 'Harar Jugol, the Fortified Town', Ethiopia, inscribed on the World Heritage List in 2006. Here the scope of management is to guarantee continuity for the traditional community, taking into account the social, cultural and economic conditions. Important also is the traditional relationship with the surrounding rural territory, which has been defined as a buffer zone. The threat in this case comes from pressure of modern developments in a new town attached to the old. Therefore, management needs to contain

Figure 2.1 Istanbul Turkey, inscribed 1985, with rebuilding of traditional housing © Author

and guide that development so as not infringe the quality of traditional life of the community in the old town (UNESCO 2006).

In the neighbouring country, Eritrea, the city of Asmara was designed by Italian colonisers and built by the Ethiopians starting from the end of the nineteenth century, until the 1940s. Subsequently, it was colonised by the British and then by Ethiopians, until independence from the 1990s. Here, the population has built up an identity with the modernist city, but there is a lack of understanding and incentives for a proper maintenance strategy. Therefore, even though the city has retained its original urban fabric reasonably well, the buildings are not in good state of conservation. Therefore, in Asmara, the question is about the development of that community's awareness through a Steering Committee. This Committee would involve the principal stakeholders and initiate a process of capacity building, aiming to learn to appreciate the significance of the existing building stock. This must be based on appropriate management strategies that take into account the building up of a socio-economic development that supports the cultural heritage assets of the place within its regional context.[4]

In Azerbaijan, the historic town of Baku was destroyed by an earthquake in the eighteenth century. It was rebuilt in the nineteenth century on the old foundations, maintaining the structure of the urban fabric and its archaeological stratigraphy. The old town (Icheri Sheher) was inscribed on the World Heritage List in 2000: 'Walled City of Baku with the Shirvanshah's Palace and Maiden Tower' (UNESCO 2000). Becoming more wealthy with the petroleum industry, the original inhabitants gradually moved out and Icheri Sheher was inhabited by immigrants from rural areas (UNESCO 1985). Unfortunately, the new inhabitants had neither an interest nor the means to care for the old town. In 2007, responding to development pressures, the President of the Republic established a new administration specifically for the historic town, distinct from the municipal authority. The problem here was that the management regime was limited to the protected area and did not consider the metropolitan context. Consequently, the historic urban landscape of Baku surrounding the old town, mainly dating from developments in the nineteenth and early twentieth centuries, has become subject to new ambitious creations and high-rise buildings, reflecting ideas from fast-developing places such as Dubai.

In the United Arab Emirates, cities like Dubai and Sharjah have become symbols of modern development expressed in ever more fantastic architectures. In both places, the traditional habitat has mostly disappeared as a result of such developments. However, there are currently ambitions to find something to nominate to the UNESCO List. The old port of Dubai has already been nominated but not yet approved, and Sharjah is also interested (UNESCO 2014). Here the ruling Sheikh has taken the initiative to re-establish the traditional habitat, recovering the remaining fragments and regenerating new building on the same typological patterns (Figure 2.2). Some of these buildings could be offered for residential use with the necessary service structures and small-scale commercial activities. The scope is to establish a reasonably large area and also replace some of the surrounding high-rise buildings, lowering their height and building more suitable structures so as to guarantee a proper scale for the traditional-type habitat.[5]

Figure 2.2 Sharjah. Heart of Sharjah. UAE, rehabilitated houses in the old city © Author

Japan has also been subject to rapid modern development following the destruction during World War II. Here, a number of traditional villages and traditional urban areas have been legally protected from the 1970s. However, larger urban areas have often lost their traditional fabric. Nevertheless, it is interesting to look at the example of Kyoto, where some areas are on the UNESCO List: Historic Monuments of Ancient Kyoto (Kyoto, Uji and Otsu Cities) (2010). Even here, the tendency has been to introduce modern high-rise buildings. However, starting from the 1990s, as a result of intense work initiated by culturally sensitive and qualified conservationists, it has been possible to build up a heritage community that is starting to appreciate the traditional habitat. Consequently, traditional buildings are now being rehabilitated and restored, and the aim is to lower the tall buildings to a more 'human scale' (Figure 2.3). An inspiration for this type of action is seen in the 'Machinami Charter' (*Charter for the Conservation of Historic Towns and Settlements of Japan*), adopted by the Japanese ICOMOS Committee in 2000.

The Machinami Charter states that the aim of conserving historic towns is not only to save groups of houses and the surrounding landscape as material objects, but also to attempt to reconstruct the relationships between the daily life of the residents, the houses and the surrounding settings. The aim is to consider the historic area, in its tangible and intangible elements, its physical and spiritual aspects, creating a 'bond of spirits' with a view to an appropriate culturally sensitive and sustainable economic development. We can understand culture, as stated above, as both the generator and a product of development within the evolving framework of the economy of a community (Jokilehto 2012).

In this respect, I reflect on the term 'gross national happiness', coined by the King of Bhutan in 1972 to signal his commitment to building an economy that would serve Bhutan's unique culture based on Buddhist spiritual values (Centre for Bhutan Studies and GNH Research 2015). Since then, the term has gained international recognition, being associated with various parameters of life quality, including cultural, social and environmental, as well as economic well-being. The Bhutanese people have been able to retain their traditional continuity better than most in our rapidly globalising world. Traditional crafts still exist, and the Buddhist script is still taken as the guidance for renovation of the existing historic buildings and even for new construction (Figure 2.4). This continuity is, however, at risk; even Bhutan is part of modernity. Much of its academic training is taken abroad, and the link with the local community remains fragile. In the country, there are rules to reflect tradition in costumes as well as in the built environment. Even the new Paro airport building is built in traditional style. However, there is a particular quality in the authentic tradition represented by genuine vernacular heritage, still common in the rural areas and visible even in the urban centre of the capital city, Thimphu. It is not the same as reinterpreting traditional forms in steel and concrete. The question can, indeed, be raised: what are the limits? We can also ask whether this reinterpretation in modern building stock is not contributing to some sort of debasement of the genuine traditions? At the same time, it has helped to provide some guidelines for modern buildings.

Figure 2.3 Kyoto, Japan: a) protected area (above) and b) community event in risk
preparedness (below) © Author

Building capacity for conservation

The development of legal instruments for safeguarding Bhutan's cultural heritage is currently underway. In this regard, we can note that such a law necessarily provides only a framework and will need to be accompanied with clear guidelines, integrated with instructions for the interpretation of the law in different cases. The problem often is that a law alone does not protect. It is effective only if it is well interpreted and if there is due awareness, not only by government officials and administrators, but also by practitioners and the local communities. Consequently, education and awareness raising are an essential tool to accompany law, in a word: appropriate capacity building, taking into account the legal, institutional, professional and community aspects. This is particularly important in Bhutan, where the people really still form the 'heritage community', as discussed above in relation to the Faro Convention. The question is about management of traditional continuity, accommodating the necessary changes without losing the qualities of historical authenticity or integrity.

It is in this context that one also understands the notion of 'capacity building', which has become a key phrase, complementing the earlier emphasis on education and training. In 2011, the World Heritage Committee adopted the strategy for capacity building (UNESCO 2011), the implementation of which is coordinated by ICCROM in collaboration with the other advisory bodies, ICOMOS and IUCN (ICCROM *et al.* 2014). The strategy aims at broadening the notion of training to capacity building. The aim is not only to build up human resources, but also to take into account administrative and institutional cooperation, as well as legal and regulatory frameworks and applicability. The aim of this World Heritage initiative is to organise a series of capacity building workshops engaged with a regional audience. The participants would represent different countries, and on their return would become initiators of relevant activities in their own region. An example is in Georgia, where ICCROM is collaborating with the national authorities to develop training curricula and to discuss the legal and administrative frameworks.

Through its Architectural-Archaeological Tangible Heritage in the Arab Region (ATHAR) Centre in Sharjah, United Arab Emirates, ICCROM has collaborated within the Arab region. An example is the capacity building programme launched in Yemen, in collaboration with the Social Fund for Development and other ministries and universities. The aim was to survey existing training at Yemeni universities, comparing with the programmes in neighbouring countries. The scope is to, first, improve training and, then, also involve professionals and administrations. Unfortunately, the political situation in the country has not allowed this to be completed as planned, but positive progress is being made in Asia and the Pacific and in Africa (ICCROM *et al.* 2014, 2015).

The importance of the role of community was not initially appreciated in the World Heritage context. However, with time it has become abundantly clear that conservation of heritage is not feasible without the involvement of the

Figure 2.4 Bhutan, a) Punakha Dzong, former government seat (above), and b) Chhungney
Guemba, mountain monastery (below) © Author

community. This is now accepted by the World Heritage Committee, as declared
in the World Heritage Kyoto Vision of 2012, noted above, and in the recent
COMPACT report on 'Engaging local communities in World Heritage' (ICCROM
et al. 2014: 24–5). One should remember, however, that the community must
include all relevant stakeholders, each with the proper role. There is need for the

necessary professional and technical knowledge and expertise, involving and sustaining the property owners and authorities in a process aiming to understand the significance of the place and establish the socio-economic conditions for the continuity of its life.

The basic message here is that the conservation and regeneration of a place must start from the recognition of its significance. And, as indicated in the 1976 UNESCO *Recommendation*, the significance is based on the recognition of human interaction as much as the material built heritage, the spatial organization and the surroundings. This is not always a simple task, and to be effective the management system must take into account the specificity of the place. A management plan that is good for one place may not be suitable for another. Indeed, as is seen above, the situations vary from one place to another. Consequently, also, the scope of management must be designed properly. It is *not* a good idea to invite an outside expert to prepare a management plan without communicating with and involving the community. However, it can also be risky to simply expect the local community to take all the necessary initiatives in the appropriate direction.

Even when the traditional community still exists, as in the case of Harar Jugol, it can be fragile and can be easily lost as happened in Istanbul and in Dubai. When the traditional community does exist it needs support and guidance by qualified persons. A properly trained and educated person should be able to communicate with the community. Without such capability, this person cannot be considered an expert. Creating a 'heritage community', as proposed in the 2005 Faro Convention, is based on a learning process which has its requirements and implications. When a place is recognised for its heritage significance, it is necessary to educate and guide the community, involving all relevant stakeholders in order to guarantee the conditions for the continuity of its heritage significance within the changing and globalising world.

In 1993, in Colombo, the ICOMOS General Assembly adopted the *Guidelines on Education and Training in the Conservation of Monuments, Ensembles and Sites*, prepared by the ICOMOS International Training Committee, CIF (ICOMOS CIF 1993). The guidelines note that: 'Conservation works should only be entrusted to persons competent in these specialist activities.' Education and training for conservation should produce a range of professionals, conservationists who are able 'to cope with specific tasks, such as analyse and understand the monuments and sites in their setting, understand and diagnose the causes of decay, and know, understand and apply conservation theory and internationally adopted guidelines' (ibid.: 43).

Twenty years later, in 2013, ICOMOS CIF drafted another document: *Principles for Capacity Building through Education and Training in Safeguarding and Integrated Conservation of the Cultural Heritage* (ICOMOS CIF 2013). This document is not intended to replace the 1993 *Guidelines*, but to update its message for capacity building, as a framework document providing overall guidance for safeguarding processes. In this context, the heritage community, as intended in the 2005 Faro Convention, certainly has a primary place. However, the community includes many different stakeholders: building owners and administrators, as well as specialised conservationists in different fields and many others. To work with communities does not mean forgetting professional knowledge. Heritage is heritage when recognised as such; taking care of heritage requires knowledge. Knowledge is based on learning. Learning is the scope of capacity building.

Notes

1 'The Kyoto Vision', adopted by the closing event of the celebrations of the 40th anniversary of the World Heritage Convention, Kyoto, December 2012.
2 In 1992, the World Heritage Committee adopted the notion of 'cultural landscape' as a new category eligible for inscription to the UNESCO List. In 2004, INTACH adopted the Charter for the Conservation of Unprotected Architectural Heritage and Sites in India.
3 Stakeholder theory, in reference to management, has been elaborated by R. Edward Freeman in his *Strategic Management: A Stakeholder Approach*, first published in 1984 (2010). The notion has become common currency also in the conservation field, though rarely in formal charters.
4 The colonial capital of Eritrea was planned and built under the direction of Italian architects at the time of Mussolini, in the 1920s and 1930s. The city of Asmara is proposed for nomination to the World Heritage List.
5 The Sheikh of Sharjah has given major attention to the respect and development of culture and traditions: details of the Heart of Sharjah project are online at http://heartofsharjah.ae/heart-of-sharjah-project.html

References

Bonelli, R. (ed.) 1978. *Scritti Rinascimentali*. Milan: Il Profilo
Centre for Bhutan Studies and GNH Research 2015. *Gross National Happiness.* Centre for Bhutan Studies and GNH Research [online]. Available at http://www.grossnationalhappiness.com/ [accessed 3 August 2015]
Council of Europe 1954. *European Cultural Convention.* Council of Europe [online]. Available at http://conventions.coe.int/Treaty/EN/Treaties/Html/018.htm [accessed 6 August 2015]
Council of Europe 2005. *Framework Convention on the Value of Cultural Heritage for Society.* Council of Europe [online]. Available at http://www.coe.int/en/web/conventions/full-list/-/conventions/treaty/199 [accessed 6 August 2015]
Dewey, J. 1934. *Art as Experience* (reprinted 2005). Harmondsworth: Penguin
Freeman, R.E. 2010. *Strategic Management: A Stakeholder Approach.* Cambridge: Cambridge University Press
Gauchet, M. 1985. *Le désenchantement du monde*. Paris: Gallimard
ICOMOS CIF 1993. *Guidelines on Education and Training in the Conservation of Monuments, Ensembles and Sites* [pdf]. ICOMOS [online]. Available at http://www.icomos.org/charters/education-e.pdf [accessed 5 August 2015]
ICOMOS CIF 2013. *Principles for Capacity Building through Education and Training in Safeguarding and Integrated Conservation of the Cultural Heritage* [pdf]. ICOMOS CIF [online]. Available at http://cif.icomos.org/pdf_docs/CIF%20Meetings/Guidelines/ICOMOS_CIF_PrinciplesCapacity_EN_20130930.pdf [accessed 5 August 2015]
ICOMOS 2014. *Florence Declaration on Heritage and Landscape as Human Values.* ICOMOS [online]. Available at http://www.icomos.org/images/DOCUMENTS/

Secretariat/2015/GA_2014_results/GA2014_Symposium_FlorenceDeclaration_EN_
final_20150318.pdf [accessed 6 August 2015]

ICCROM, ICOMOS, IUCN 2014. *World Heritage Capacity Building Newsletter No. 4*
[online]. Available at http://www.iccrom.org/world-heritage-capacity-building-newslet-
ter-now-out/ [accessed 3 August 2015]

ICCROM, ICOMOS, IUCN 2015. *World Heritage Capacity Building Newsletter No. 5*
[online]. Available at http://www.iccrom.org/world-heritage-capacity-building-newslet-
ter-no-5/ [accessed 3 August 2015]

Jokilehto, J. 2012. 'Culture as a factor of development', *Territori della Cultura* 8. Ravello,
Italy: Centro Universitario Europeo per i Beni Culturali, 58–67

Okinawa International Forum, 2004. *Utaki in Okinawa and Sacred Places in Asia:
Community Development and Cultural Heritage.* Tokyo: Japan Foundation

Taylor, C. 1991. *The Ethics of Authenticity.* Cambridge, MA: Harvard University Press

UNESCO 1945. *Constitution of the United Nations Educational, Scientific and Cultural
Organisation.* UNESCO [online]. Available at http://portal.unesco.org/en/ev.php-URL_
ID=15244&URL_DO=DO_TOPIC&URL_SECTION=201.html [accessed 6 August
2015]

UNESCO 1972. *Convention Concerning the Protection of the World Cultural and Natural
Heritage* [pdf]. Paris: UNESCO [online]. Available at http://whc.unesco.org/archive/
convention-en.pdf [accessed 5 August 2015]

UNESCO 1976. *Recommendation Concerning the Safeguarding and Contemporary Role
of Historic Areas* [online]. Available at http://portal.unesco.org/en/ev.php-URL_
ID=13133&URL_DO=DO_TOPIC&URL_SECTION=201.html [accessed 6 August 2015]

UNESCO 1985. *Historical Areas of Istanbul.* UNESCO [online]. Available at http://whc.
unesco.org/en/list/356 [accessed 3 August 2015]

UNESCO 2000. *Walled City of Baku with the Shirvanshah's Palace and Maiden Tower.*
UNESCO [online]. Available at http://whc.unesco.org/en/list/958 [accessed 5 August 2015]

UNESCO 2003. *Convention for the Safeguarding of the Intangible Cultural Heritage.*
UNESCO [online]. Available at http://portal.unesco.org/en/ev.php-URL_
ID=17716&URL_DO=DO_TOPIC&URL_SECTION=201.html [accessed 3 August 2015]

UNESCO 2005. *Convention on the Protection and Promotion of the Diversity of Cultural
Expressions.* UNESCO [online]. Available at http://portal.unesco.org/en/ev.php-URL_
ID=31038&URL_DO=DO_TOPIC&URL_SECTION=201.html [accessed 3 August 2015]

UNESCO 2006. *Harar Jugol, the Fortified Historic Town.* UNESCO [online]. Available
at http://whc.unesco.org/en/list/1189/ [accessed 3 August 2015]

UNESCO 2010. *Historic Monuments of Ancient Kyoto (Kyoto, Uji and Otsu Cities).*
UNESCO [online]. Available at http://whc.unesco.org/en/list/688 [accessed 3 August
2015]

UNESCO 2011. *Presentation and Adoption of the World Heritage Capacity Building
Strategy* [pdf]. Paris, France UNESCO [online]. Available at http://whc.unesco.org/
archive/2011/whc11-35com-9Be.pdf [accessed 6 August 2015]

UNESCO 2013. *Operational Guidelines for the Implementation of the World Heritage
Convention* [pdf]. Paris.: UNESCO [online]. Available at http://whc.unesco.org/archive/
opguide13-en.pdf [accessed 3 August 2015]

UNESCO 2014. 'Operational Directives for the Implementation of the Convention for the
Safeguarding of the Intangible Cultural Heritage' [pdf]. In UNESCO *Basic Texts of the
2003 Convention for the Safeguarding of the Intangible Cultural Heritage.* Paris
[online]. Available at http://unesdoc.unesco.org/images/0023/002305/230504e.pdf
[accessed 3 August 2015]

3 People-centred approaches

Engaging communities and developing capacities for managing heritage

*Gamini Wijesuriya, Jane Thompson
and Sarah Court*

Introduction

There are changing demands on, and expectations of, cultural heritage in society and a need for approaches that are built on greater consensus and collaboration to ensure objectives are met in a sustainable way. Improving the relevance and effectiveness of contributions of those already involved in conservation and management of heritage, as well as facilitating the engagement of new audiences, has become a priority for many countries in the twenty-first century. The wellbeing of people around the world has become a major concern in all spheres, so the heritage sector has no option but to participate in this wider context and contribute to a better future for people and our planet, or it risks being sidelined. It is in this context that engaging people in the heritage process has become one of the most dominant issues being debated in the sector at present.

The potential of heritage to play an active role in the lives of communities and to bring benefits to people is increasingly widely recognised. Experience in the field analysed with this lens is demonstrating that, on one hand, heritage is meaningful to society and, on the other, gaining society's support is significantly beneficial to ongoing use and protection (Galla 2012). Engaging communities is about strengthening their ability to participate meaningfully in the process of making conservation and management decisions for themselves and their heritage and in the process of implementing them.

On the basis of a discussion of the importance of 'people' in relation to heritage, this chapter will summarise some of the contributions ICCROM has made as an intergovernmental organisation over the last two decades to ensure people are at the heart of heritage conservation and management through a range of policy and capacity building activities for the heritage sector. Progress has been achieved through fostering institutional partnerships and intersectorial awareness and, above all, by capturing inputs from the vast and ever more connected community of heritage practitioners who benefit from and contribute to ICCROM's programmes. It has addressed what was essentially a fundamental gap in the modern conservation discourse that had evolved in the Western world – the omission of communities and their role (Wijesuriya 2008) – and the negative repercussions this was having on heritage management practice, particularly with the trend for more complex multiple-ownership and multiple-use heritage typologies.

ICCROM's work has shown the importance of improving understanding of the sector's management systems, strengths and shortcomings, and comprehension of where heritage capacities reside and the role of people to forge positive change. A greater ability to identify when and where capacity needs creating or developing – either within local communities or inside the institutional frameworks and among the practitioners of the heritage sector itself – and then frame the capacity development process as a form of people-centred change, is central to achieving an upward spiral of continuous improvement in heritage management practice, with the benefits this will have for society today and in the future.

Heritage and society, an international perspective

In 1999, then Director General of UNESCO, Koïchiro Matsuura stated that, 'Without the understanding and support of the public at large, without the respect and daily care of the local communities, which are the true custodians of World Heritage, no amount of funds or army of experts will suffice in protecting the sites' (UNESCO World Heritage Centre 1999). His words reflected a growing awareness in the international community of inadequacies in many management and conservation approaches to cultural heritage.

It has been a long journey from those words to the adoption of the *Policy for the Integration of a Sustainable Development Perspective into the Processes of the World Heritage Convention* in 2015 by the 20th General Assembly of the World Heritage Convention (UNESCO World Heritage Centre 2015). This journey to increase engagement of communities in World Heritage processes, starting from nomination and management processes, has been made up of a series of milestones. With the adoption in 2007 of 'Communities' as one of the five strategic objectives for World Heritage, UNESCO gave primary importance to engaging communities within all World Heritage processes. This emphasis was consolidated further with the 40th anniversary of the World Heritage Convention in 2012 being dedicated to 'Sustainable Development and the Role of Local Communities'.

This entire process can be considered an attempt to strengthen an aspect of the *Convention* that was forgotten for nearly 40 years, that of heritage having 'a function in the life of the community' (UNESCO 1972). Over time the World Heritage Committee and the broader heritage sector has learnt to build on the diversity of experiences worldwide and draw on the broader sustainable development paradigm to highlight the need to consider reciprocal benefits for both heritage and communities in heritage management approaches. This shift has given greater importance to heritage communities and heritage 'users' (those in the present, not just in the future). Moreover, it has recognised that use patterns and cultural practices are often central to the cultural values of a place.

The stimulus for this shift came from diverse cultural heritage approaches that had been off the radar of the international community for too long as the following section illustrates. ICCROM, as one of the three Advisory Bodies to the World Heritage Committee and, in particular, as the priority partner for training, continually reaches out to the field for capacity developing activities and research, and has played a significant role in promoting the 'people' factor in policy work and practice promoted by the international heritage community.

Learning from 'living heritage'

Heritage has been created *by* people and it has been created *for* people and, without a doubt, it has been created to support the wellbeing of people. In a world where people and their livelihoods are the focus of much international attention, the heritage sector can no longer focus on heritage alone but also needs to look at how it can contribute to, and benefit from, broader socio-economic and environmental wellbeing. For these reasons, people are increasingly considered as a core part of the heritage management equation.

Postgraduate research carried out in York in the early 1990s by one of the authors (Wijesuriya 1993) compared and contrasted – and even justified – restoration and reconstruction work on Buddhist monuments as cultural heritage belonging to living religious traditions in Sri Lanka. This experience led the author to test some of the theoretical perspectives at a later stage when one of the most sacred places in Sri Lanka, the Temple of the Tooth Relic (which is also a World Heritage property), was bombed by terrorists in 1998 (Figure 3.1). As living heritage, this opportunity allowed the author to reflect on the work carried out in York, yet within a process that was completely driven by the communities and custodians of the place. There is an assumption that heritage processes are all led by professionals and that community involvement is somehow being shoe-horned in – in this way community contributions are allowed to embellish professional efforts. However, the Temple of the Tooth Relic experience demonstrated a different paradigm: heritage is not heritage, it is a part of community life and it is therefore very natural for that community to care for it. It was shown that when professionals try to insert themselves into *this* context, things can go wrong (Wijesuriya 2001, 2005, 2007).

This notion of living heritage, bottom-up approaches, the diverse definitions of heritage held by the custodians of that heritage and the benefits they gain were recorded in a research paper presented at an ICCROM experts' meeting which was later to become the backbone of ICCROM's newly launched Living Heritage Sites programme. Further research on living heritage (Wijesuriya 2015b) has recognised the influence of heritage on the contemporary life of people and how it can improve their quality of life, focusing on both past and present and enhancing the value of cultural products. It has highlighted the need to respect diversity, to improve relationships between heritage and people, and to recognise the living dimensions of heritage. It has also recognised the fact that communities have capacities and assets that outlast political or professional structures and complement specialist knowledge and skills in managing heritage. It is in this context, that the custodianship of communities for the long-term care of their heritage (with their traditional knowledge systems) and the need to respect people's voices

Figure 3.1 The restoration of the Temple of the Tooth Relic, Sri Lanka, raised
questions of who should lead the heritage processes © Gamini
Wijesuriya

in the conservation and management of heritage has been recognised. These inter-linkages between people, their wellbeing and their heritage became the focus of ICCROM's Living Heritage Sites programme. This focus extended towards the more active engagement of people (most usefully worked with in the form of communities) in all aspects of heritage processes and with the aim of harnessing their capacities in order to offer long-term conservation and co-management for the good of the heritage and for the good of the community. Conversely, there are many examples that illustrate the negative impacts that can occur when heritage is divorced from society by an imbalanced management system (e.g. Taruvinga and Ndoro 2003).

It is also becoming necessary to take into consideration the impact of globalisation on living environments, such as historic urban centres and cultural landscapes, as well as the links between heritage and the sustainable development of society. In many ways, heritage that continues to perform functions for society has not faced that divorce from present society, the isolation that the 'museumification' process that many Western management systems have created, and which can create negative repercussions for the entire management and conservation approach. A living heritage approach – or a heritage in use approach – favours:

- a more 'people-up' approach;
- benefits being gained by both heritage and people;
- continuity and sharing of knowledge and resources;
- a more direct connection to livelihoods and thus social, economic and environmental values;
- primary stakeholders being central to decision-making, not just being placed in a secondary role;
- the engagement of the primary holders of cultural values – the connected community – being unavoidable.

Wijesuriya 2015a

The overall lesson to be drawn is that communities cannot just be considered as another category of stakeholders, but must be a *sine qua non* within the heritage discourse. This recognition has led to new approaches being explored, such as living heritage or, more broadly, people-centred approaches and to question much of what has become established practice in the international heritage sector. These questions have included: where do heritage capacities reside and how can they be created or reinforced? What constitutes satisfactory 'community engagement' for all heritage typologies? How is it possible to better understand the role of communities in capacity building? How can common ground be located between diverse heritage management systems and approaches?

Learning from living heritage: ICCROM capacity development

This theoretical basis has influenced ICCROM's capacity building initiatives, where programmes have long been organised that included a strong emphasis on community engagement. The Living Heritage Sites programme that ran between 2003 and 2008 was an early example (Wijesuriya 2015a; Stovel *et al.* 2005) and the rationale behind the programme was to emphasise the living dimensions of heritage sites: their recognition and relevance to contemporary life, including benefits and people's interests and capacity to engage in continuous care as true and long-term custodians of these sites. It was surmised that 'continuity' of heritage for the purpose for which it was created was inseparable from the community connected to it and the benefits they derived from it, which, in turn, makes heritage significant to the wellbeing of the people. Retaining the living dimensions of heritage that contain and support diverse socio-cultural activities was considered as important as the material fabric. It concluded that heritage conservation is about creatively managing continuity, which is related to the growing notion of managing change: indeed continuity is inherent to change.

This focus on people in the activities of the Living Heritage Sites programme highlighted how essential it is to change our conventional approaches to the conservation and management of heritage and how these should be driven by people-led or 'people-up' processes (Figure 3.2; Wijesuriya *et al.* 2013: 24–8). Indeed, it should be noted how the Living Heritage approach managed to address some of the gaps in other approaches to heritage management – such as diversity, context dependency and community in decision-making processes – in defining, conserving and managing heritage. This demonstrated that while living heritage is proving to be a useful framework for conservation globally, where there is still a clear living heritage tradition with continuity of use (e.g. religious buildings, urban landscapes, even such heritage as the London underground, etc.), it is an approach that can be readily adopted and adapted for heritage in other contexts. In particular, this is true where communities have been cut off from their heritage by modern heritage management systems yet where attempts are being made to reinstate the heritage/community relationship.

This recognition that there were many more typologies of heritage that could benefit from some of the approaches developed during the experience of the Living Heritage Sites programme led to the launch of ICCROM's more recent programme on 'Promoting People-Centred Approaches to Conservation', one of the five priority programmes for the period 2012–17. Engaging communities in conservation and management processes is considered a key component within people-centred approaches: as already mentioned above, it is about strengthening their ability to participate meaningfully in the process of making conservation and management decisions for themselves and their heritage. As involving communities can be a real challenge at many heritage places, and considering the challenges of transmitting community concerns to heritage practitioners, the programme was evolved with a view to promoting the engagement of communities in the conservation and management of heritage.

Input from diverse practitioners and heritage organisations was the basis for initial research in this area (Court and Wijesuriya 2015) and, in turn, the first capacity building activity of this programme: the 2015 short course on 'Promoting People-Centred Approaches: Engaging Communities in the Conservation of

Conventional (fabric-focused) conservation approach

Values-based approach to conservation

People-centred approaches: living heritage

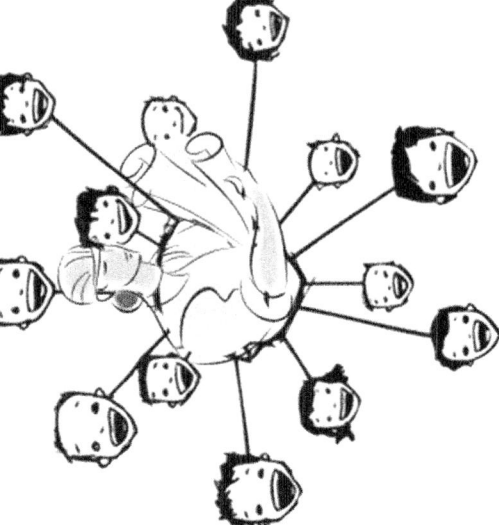

Figure 3.2 Evolving approaches to heritage management © Gamini Wijesuriya

Nature and Culture' (PCA15). The course was based on the premise that, while there is no simple recipe for community engagement, there are many examples that can be explored to understand the range of possible approaches and to inspire adapted approaches elsewhere. PCA15 was, therefore, an opportunity to bring practitioners together to share their experiences and learn from others, so as to move practice forward by providing them with the necessary knowledge and tools to work more effectively with communities through existing management systems. The course was an opportunity to create a forum for participants to share their experiences from both the cultural and natural heritage sectors, learning from each other and other heritage practitioners who are actively involved with communities. Importantly, over half of the course programme was dedicated to the participants presenting their own work in this area or visiting sites, allowing discussion to remain firmly grounded in the reality of working in the field. Case study sites were chosen from around the Bay of Naples (Italy), where community groups of different types are actively engaged in the management and conservation of their heritage in a range of ways.

One case study explored within the course raises many relevant issues for this chapter. The PCA15 participants were shown round the neighbourhood of Sanità, in the centre of Naples, by a community group (Figure 3.3). Despite a wealth of heritage assets and being located in the heart of the World Heritage property of the Historic Centre of Naples, it has long been an area associated with urban decay and major social problems. However, significant changes have been seen since a series of cooperatives and associations were formed by groups of local residents, each with the aim of improving the lives of the local community. Many of these have been related to cultural heritage; in particular, La Paranza Cooperative was created in 2006 by a group of young residents: initially providing heritage walking tours of their neighbourhood, they later took over management of the catacombs that lie under this area, providing full visitor management, site presentation and interpretation. Their efforts now extend to more elements of the neighbourhood's rich tangible and intangible heritage and are all aimed at promoting public access and enjoyment of this heritage as part of process of self-development for residents (La Paranza 2016).

This group of young people has managed to improve conservation conditions while providing public access and a new identity for this urban district. Through tourism and cultural initiatives, they have managed to create jobs, raise awareness of Sanità's rich heritage and open up the neighbourhood to economic benefits from visitors. They have collaborated with a number of other local groups, providing training and employment opportunities to disadvantaged young people while also creating a technical community for the upkeep and maintenance of the heritage (Loffredo 2013). The results of these efforts are visible in the professionally managed visits to the local heritage, with more available to the public each year, increased visitor numbers, ongoing conservation efforts and enhanced cultural values at a local, national and international level. The work underway at Rione Sanità illustrates how local community-led efforts are transforming a neighbourhood, with benefits for both the residents and the heritage. Broad participation in

Figure 3.3 Participants of ICCROM's first course on 'Promoting People-Centred Approaches' visited case studies around the Bay of Naples (Italy) where communities have a key role to play in heritage – such as in Rione Sanità shown here – in order to root discussion in the reality of managing heritage © ICCROM

cultural heritage and the subsequent increase in tourism has provided job opportunities for individuals and income for small businesses – but not only that: it has ensured that the heritage has been reappropriated by the community, it is cared for and accessible to the public (Court and Thompson forthcoming 2016).

None of the case studies explored by the PCA15 participants were framed as a 'model' of community participation, but analysis of such real-life examples might help identify some key capacities required for successful management. In the Sanità neighbourhood, groups from within the local community have found a way to take on the daily management of their heritage from the heritage authorities, using it as a tool for providing socio-economic and cultural benefits for residents, creating tourism in a formerly no-go area and ensuring that the heritage is cared for and protected. It is significant how that broader agenda has enriched their management of what elsewhere would be solely a heritage destination, the catacombs. Their management capacities, together with their ability to reinforce the identity and values of the places they recount to visitors, mean it is becoming one of Naples' most popular visitor destinations. Moreover, the Rione Sanità experience has, in part, inspired 2015 legislative changes in Italy that allow management of cultural assets to be devolved to non-profit associations.

However, in this very positive climate, the relative absence of the heritage institutions is of note and raises concerns about the balance of stakeholders contributing to heritage management in this case. As a single case study, it reflects a wider difficulty in the heritage sector whereby the primary and often public national heritage institutions, often solely preoccupied with preservation of the physical testimonies of the past and without the capacity to address all needs, are unable to tap into the additional resources that a local community can bring. In this – as in many cases – the local community has existing capacities that can be used to gain benefits for the heritage and for themselves. The heritage management system can potentially draw benefits from capacities being sourced more widely – in particular, unlocking community participation and thereby the reciprocal benefits. Analysis of such examples should inform capacity building strategies in order to ensure that initiatives proposed for such heritage places recognise existing capacities and note gaps that need filling. In some cases, it should be recognised that institutional frameworks and heritage practitioners need their capacities to be developed in order to be able to work with communities and tap into their contribution.

An international framework for changing audiences and new learning environments

Although genuine community engagement in heritage remains a huge challenge, ICCROM's core activity, practitioner training over several decades, confirms that bringing together heritage specialists who work in the field in diverse parts of the world not only allows the sharing of lessons learnt, but also ensures positive experiences are adapted and reapplied elsewhere. Indeed, ICCROM's role as one of the three advisory bodies to the World Heritage Committee and, in particular, as the priority partner for training, has made it well placed to work with all interested parties to ensure knowledge gains by researchers and advances by practitioners are constantly mutually enriched. This, in turn, has fed back into international policy work; the World Heritage Capacity Building Strategy of 2011 is a case in point.

The 2011 World Heritage Capacity Building Strategy (UNESCO World Heritage Centre 2011) arose directly from the need to step beyond conventional training of practitioners and embrace an approach capable of reaching diverse and growing audiences, also through a much broader range of learning environments that links heritage into their broader context, including communities (Table 3.1). This is a vital shift from the idea of knowledge transfer (training) to knowledge acquisition in order to reach new and diverse audiences (including those who may not perceive their need). The Capacity Building Strategy acknowledges that the ability to create or reinforce heritage capacities resides in human beings and that capacity building needs to work with and through people.

The work of the strategy also laid a strong foundation for joint activities between cultural and natural heritage sectors which had previously been operated in isolation, despite the fact that the *Convention* had been intended in a different way (Larsen and Wijesuriya 2015). Such activities – led by ICCROM with other institutions IUCN and ICOMOS – further strengthened the need to drive for greater engagement of communities in heritage processes of both sectors.

Understanding the gaps in a management system to recognise the 'where' and 'who' of capacity building

Many of the experiences and advances mentioned above have been brought together and linked within the resource manual on *Managing Cultural World Heritage* (Wijesuriya *et al.* 2013: Chapter 2) led by ICCROM. This and other manuals in the series were a response by the UNESCO advisory bodies (ICOMOS, ICCROM and IUCN) and the World Heritage Centre to the lack of tools for assessing positive impacts of heritage and effective management, and harnessing them for the benefit of both heritage and society.

The manual offers a rare overview of current thinking in the heritage sector (Part 2), including an outline of sustainable development as a dominant paradigm of our time and the role cultural heritage conservation and management can play. This translates most immediately into the concern for sustaining heritage, considered as an end in itself, as part of the environmental and cultural resources that should be protected and transmitted to future generations. Perhaps more significantly for the themes of this chapter, it is also interpreted as the possible contribution that heritage and heritage conservation can make to the environmental, social and economic dimensions of sustainable development and hence broader wellbeing in society. The manual's core section (Part 4) aims to augment understanding of heritage management systems through a conceptual framework that acknowledges their complexity and diversity but also draws out characteristics common to all (Table 3.1).

Table 3.1 This identification of where capacities reside, related audiences and learning areas in the heritage sector forms the basis for the World Heritage Capacity Building Strategy

Where capacities reside: target audiences for capacity building	*Principal learning areas*
Practitioners (including individuals and groups who directly intervene in the conservation and management of World Heritage properties)	- Implementation of the Convention (Tentative Lists, nomination, etc.) - Conservation and management issues: planning, implementation and monitoring - Technical and scientific issues - Resource utilisation and management
Institutions (including state party heritage organisations, NGOs, the World Heritage Committee, advisory bodies and other institutions that have a responsibility for the enabling environment for management and conservation)	- Policy-making for learning areas mentioned above - Legislative issues - Institutional frameworks/issues (governance, decentralisation) - Financial issues - Human resources - Knowledge
Communities and networks (including local communities living on or near properties, as well as the larger networks that nurture them)	- Reciprocal benefits and linking with sustainable development and communities - Stewardship - Communication/interpretation

Source: Wijesuriya *et al.* 2013: 51

The framework places emphasis on results with a dual focus on routine deliverables (outputs) and much broader objectives that are the driver of the management system (outcomes). This is done specifically to draw attention to the multiple benefits many heritage conservation actions can bring. An example is the immediate result of diversification of visitor itineraries on site conservation and visitor enjoyment paired with the broader repercussions this change might have on the local economy, as new audiences and repeat visitors are attracted to the area and stay longer. In other words, the framework attempts to encourage heritage practitioners to frame every specific conservation process as a contribution to society, not only in social but also in economic and environmental terms. The choice of framing 'improvements to the management system' as a result in itself was motivated by the need to encourage an ability to ensure past and current experience informs future management approaches, with inputs drawn from feedback from both within and without the management system.

This framework is a basis for understanding, defining and assessing management systems, assisting not only practitioners, but also policy-makers and communities, in better defining issues and identifying solutions in all areas of management of cultural heritage as diverse as historic cities, cultural landscapes, individual monuments or archaeological sites. It constitutes a tool for identifying gaps in different components and knowledge areas of a given management system (Wijesuriya and De Caro 2012). Significantly for the themes of this chapter, an analytical template accompanies the common management system framework to build awareness and understanding of how heritage management can depend on contributions from multiple management systems – however formal or informal and with many outside the heritage sphere – and how this must inform future practice (Figure 3.4).

An Appendix explores management planning as perhaps the only management tool in the heritage sector that aims to improve all areas of the management system (Wijesuriya *et al.* 2013: Appendix A1). Particular attention is given to the process whereby consultation processes involving all interested parties lead to the identification of factors that can create both positive and negative impacts on the heritage and influence the safeguarding and enjoyment of the heritage and the wider benefits it can deliver society.

Understanding and promoting community engagement in all areas of heritage management

The authors suggest that this research and the multiple resources explored above could be usefully united to improve the relevance of the heritage discourse and forge lasting improvements to existing heritage management systems by making sure strategies for cultural and natural heritage:

- draw on the management system framework to identify gaps and needs in current management, but also where contributions to management come from and potential forms of new support;
- reinforce existing or create new heritage capacities which reside among practitioners, within institutional frameworks and among communities and networks through learning environments that match each audience. This would potentially improve the relevance and effectiveness of contributions of those *already* involved in conservation and management of heritage, while also facilitating the involvement of new audiences who are currently left out;
- ensure that reciprocal benefits are obtained by both communities and heritage;
- achieve this through people-centred change, which entails working with groups of individuals and communities and customising learning environments to their needs.

It is proposed that the result over time should be stronger organisational frameworks and interfaces between heritage and the wider context, with communities playing a key role in both spheres free of boundaries. Only in this way will reciprocal benefits for both heritage and communities be harnessed and will heritage be assured a more central role in society in the present, not just the future.

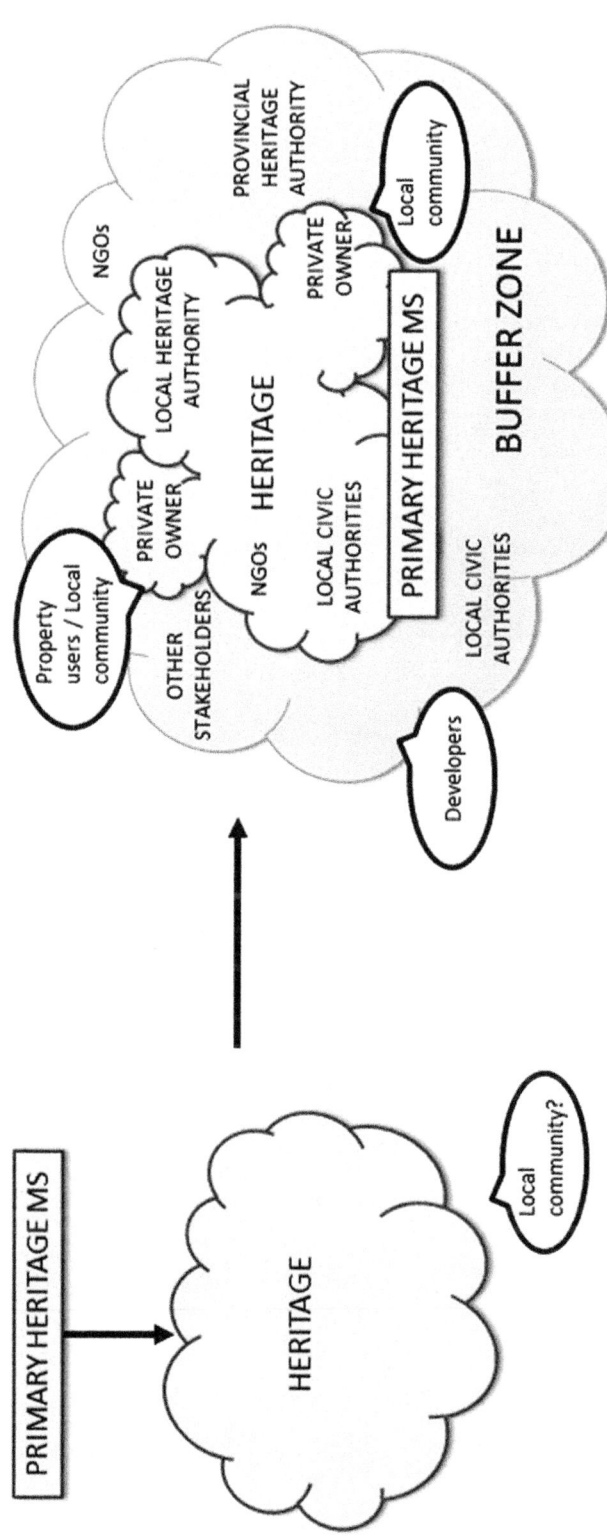

Figure 3.4 One primary management system (MS in the diagram) working with contributions from several others © Gamini Wijesuriya and Jane Thompson, ICCROM course presentation (CBH12)

Conclusions

This chapter has offered an overview of heritage sector responses and, in particular, ICCROM's response to the increasing recognition that the world is changing and that heritage needs to have a function in society, contributing to human well-being and benefiting from society, with the reciprocal gains this creates further intensifying the relationship between communities and their heritage. There are changing demands on, and expectations of, cultural heritage in society and a need for approaches built on greater consensus and collaboration to ensure objectives are met in a sustainable way. ICCROM has been working at an international level to support this shift through research and other initiatives in many areas and, in particular, on heritage management systems, capacity building and the role of communities. It is hoped that this will reduce the risk of heritage practitioners shoe-horning themselves into management scenarios where their role should perhaps be secondary to that of the core heritage community, as well as reducing the instrumentalisation of community participation in heritage management for political purposes. Indeed, taking a people-centred approach is not simply a suggestion for increasing participation within a management system; it is about addressing a core part of heritage and of heritage management – the people who are connected to the heritage – and ensuring that it is an integral element of conserving that heritage.

While there is no simple recipe for meaningful community engagement, there are many examples that can be explored to understand the range of possible approaches and to inspire their adaption for use elsewhere and gradually inform future policy and improvements to heritage management frameworks. ICCROM aims to bring practitioners together to share their experiences and learn from each other precisely to this end.

References

Court, S. and Thompson, J. (forthcoming) 2016. 'Exploring community participation in the management of Italian archaeological sites'. In I. Gürsu (ed.), *Public Archaeology*. Ankara: British Institute at Ankara

Court, S. and Wijesuriya, G. 2015. *People-Centred Approaches to the Conservation of Cultural Heritage: Living Heritage* [online]. Available at http://www.iccrom.org/wp-content/uploads/PCA_Annexe-2.pdf

Galla, A. (ed.) 2012. *World Heritage: Benefits Beyond Borders*. Cambridge: Cambridge University Press

Larsen, P. and Wijesuriya, G. 2015. 'Nature–culture interlinkages in World Heritage: bridging the gap'. *World Heritage Review* 75, 4–15 [online]. Available from: http://whc.unesco.org/en/review/75/

La Paranza 2016. *About Us* [online]. Available at http://www.catacombedinapoli.it/en/about

Loffredo, A. 2013. *Noi del Rione Sanità*. Rome: Mondadori

Stovel, H., Stanley-Price, N. and Killick, R. (eds) 2005. *Conservation of Living Religious Heritage*. Rome: ICCROM [online]. Available at http://www.iccrom.org/ifrcdn/pdf/ICCROM_ICS03_ReligiousHeritage_en.pdf

Taruvinga, P. and Ndoro, W. 2003. 'The vandalism of the Domboshava rock painting site, Zimbabwe: some reflections on approaches to heritage management'. *Conservation and Management of Archaeological Sites* 6(1), 3–10

UNESCO 1972. *Convention Concerning the Protection of the World Cultural and Natural Heritage.* Paris: UNESCO [online]. Available at http://whc.unesco.org/archive/convention-en.pdf

UNESCO World Heritage Centre 1999. *Address by Mr Koïchiro Matsuura, Director-General of UNESCO at the Opening of the 23rd Session of the World Heritage Committee. DG/99/3/KM* [online]. Available at http://unesdoc.unesco.org/images/0011/001182/118260E.pdf

UNESCO World Heritage Centre 2011. *Presentation and Adoption of the World Heritage Strategy for Capacity Building. WHC-11/35.COM/9B* [online]. Available at http://whc.unesco.org/archive/2011/whc11-35com-9Be.pdf

UNESCO World Heritage Centre 2015. *Policy for the Integration of a Sustainable Development Perspective into the Processes of the World Heritage Convention. WHC-15/20.GA/INF.13* [online]. Available at http://whc.unesco.org/document/139146

Wijesuriya, G. 1993. *Restoration of Buddhist Monuments in Sri Lanka: The Case for an Archaeological Management Strategy.* Colombo: ICOMOS

Wijesuriya, G. 2001. '"Pious vandals": restoration or destruction in Sri Lanka?'. In R. Layton, P.G. Stone and J. Thomas (eds), *Destruction and Conservation of Cultural Property.* London: Routledge, 256–63

Wijesuriya, G. 2005. 'The past is in the present: perspectives in caring for Buddhist heritage sites in Sri Lanka'. In H. Stovel, N. Stanley-Price and R. Killick (eds), *Conservation of Living Religious Heritage.* Rome: ICCROM, 31–43 [online]. Available at http://www.iccrom.org/ifrcdn/pdf/ICCROM_ICS03_ReligiousHeritage_en.pdf

Wijesuriya, G. 2007. 'The restoration of the Temple of the Tooth Relic in Sri Lanka: a post-conflict cultural response to loss of identity'. In N. Stanley-Price and R. Killick (eds), *Post War Recovery and Conservation.* Rome: ICCROM, 87–98. Available at http://www.iccrom.org/ifrcdn/pdf/ICCROM_ICS06_CulturalHeritagePostwar_en.pdf

Wijesuriya, G. 2008. 'Conservation in context'. In M.S. Falser, W. Lipp and A. Tomaszewski (eds), *Conservation and Preservation: Interactions between Theory and Practice in Memoriam Alois Riegl (1858–1905).* Vienna: Edizioni Polistampa

Wijesuriya, G. 2014. 'Introducing people-centred approach to conservation of Hani Rice Terraces'. Paper presented at the International Workshop on the Sustainable Development of the Cultural Landscape of Honghe Hani Rice Terraces, 27–31 October

Wijesuriya, G. 2015a. 'Introduction'. In G. Wijesuriya and S. Lee (eds), *Asian Buddhist Heritage: Conserving the Sacred.* Rome: ICCROM, 1–10

Wijesuriya, G. 2015b. *Living Heritage: A Summary* [online]. Available at http://www.iccrom.org/wp-content/uploads/PCA_Annexe-1.pdf

Wijesuriya, G. and De Caro, S. 2012. 'Engaging communities: approaches to capacity building'. In UNESCO World Heritage Centre (ed.), *Involving Communities in World Heritage Conservation: Concepts and Actions in Asia.* Seoul: ICOMOS-KOREA

Wijesuriya, G., Thompson, J. and Young, C. 2013. *Managing Cultural World Heritage.* Paris: UNESCO World Heritage Centre [online]. Available at http://whc.unesco.org/en/news/1078

4 Everyone loves a good story
Narrative, tradition and public participation in conservation

Nigel Walter

Context

Modern conservation has succeeded in the last 50 years in establishing an impressive international, regional and local professional infrastructure for the care of historic environment. There have been many successes to celebrate, and one could be forgiven for believing that all is, and can only continue to be, well. Over that time conservation theory has matured; Kate Clark (2014: 65–6), in her recent review of current values-based heritage management in the UK, notes the significant change in methodology stemming from the *Burra Charter*'s (2000) emphasis on significance:

> Values-based management involved a fundamental intellectual shift from heritage decisions based purely on the individual expertise of the heritage professional to a more transparent process of analysis and diagnosis. Ultimately, values-based management was more than a process; it was a different way of thinking about cultural heritage.

However, in Miles Glendinning's view, all is not as well as it seems. In his history of the modern conservation movement, he characterises our times as 'the present period of apparent disorientation' (Glendinning 2013: 450) reflecting an emerging conflict between the relatively stable methodology and theory of modern conservation and fresh challenges thrown up by the changing nature of the broader culture to which conservation belongs. And despite the benefits attributed to a values-based methodology, Clark (2014: 70) also notes with regret that 'Cultural heritage is often left out of high-level strategic thinking about the environment, about culture, about cities, and about quality of life'; what is not asked is what role, if any, a values-based methodology itself might play in this cultural disenfranchisement. The question of whether significance can be defined in quite such a manner is examined by, among others, Fredheim and Khalaf (2016), who comment that the multiplicity of value typologies in conservation is indicative of a discipline incapable of self-reflection, and propose an alternative framework that is 'dialogical' (Harrison 2013) and 'symmetrical' (Schofield 2009).

We need not concern ourselves here with questions of whether our culture is experiencing a radical break between modernity and postmodernity, or merely a more modest adjustment within modernity, as argued, for example, from a Marxist perspective by Alex Callinicos (1989). What seems undeniable is that we are living in a time of significant cultural change, with the questioning of centralised forms of authority and greater calls for broader public participation in many areas of public life. Erica Avrami (2009: 177) has called for a 'new emphasis on the social processes of conservation' yet, arguably, little progress has been made in this direction. Conservation remains overwhelmingly dependent on the expert view of the 'professionals', with a profound unease at the prospect of relinquishing influence to the 'amateurs', including those communities who bear responsibility for the upkeep of many historic buildings. And yet by remaining expert-led, conservation practice makes itself an increasing oddity in a climate of greater public participation.

This chapter takes the broadly optimistic view that the wider cultural changes we are experiencing present an opportunity to address a number of hitherto unresolved issues within modern conservation, and that these tensions are not new but have been latent within conservation from the outset. After looking at some of these issues in brief overview, the chapter will focus on public participation, and how an alternative approach based on narrative might address what is becoming an increasingly pressing concern for conservation practice.

Some 'unresolveables'

This chapter is written with the benefit of some 25 years' experience in architectural practice, with a particular focus on church communities struggling with the upkeep and adaptation of historic buildings. From that community perspective, there are various tensions evident within modern conservation; among the more obvious is the way in which conservation, at least as practised in the UK, deals with change. Despite Historic England's (formerly English Heritage) helpful definition of conservation as the management of change (2008: 71), in many places within the conservation professions there remains a resistance to many aspects of change. This might be characterised, for example, in the opposition by some local authority conservation officers and some of the national and local amenity societies to the organic growth of historic buildings by extension or alteration, it making little difference whether the building be a Grade 2 listed vernacular cottage or a major Grade 1 listed medieval church. At times it seems the understanding is that all change is loss and unequivocally a threat, and should therefore be resisted as a matter of principle. We have of course witnessed much change that has not been for the good. There have been many examples of tragic losses to the historic environment, and conservation bears the scars of battles lost; for example, in post-war Europe it is undeniable that there were many examples of salvageable historic fabric swept away by crassly inappropriate modern development in the name of progress.

But here there is a fundamental distinction to be drawn between the limitations of progress-driven modernity and change as such. If the Historic England definition of conservation as the management of change is to have any meaningful application, then as well as defending against too much change, the definition would imply that there is at least some danger of allowing *too little* change. I suggest part of the reason that change presents the difficulties that it does is because the current methodology has no means of evaluating the propriety or desirability of proposed change, or indeed the quality of proposed design. Fundamentally, conservation, at least as most frequently practised, lacks an adequate account of how heritage is created, and therefore of how change can positively enhance an evolving and inclusively owned historic environment. For all its celebration of heritage past, when it comes to contemporary cultural debate conservation is largely mute and culturally barren about heritage future. This is a huge omission and leaves conservation vulnerable to attack; curatorship is not (and, of course, never has been) enough.

Ioannis Poulios (2014) helpfully addresses the issues of change to historic structures in community ownership in the context of the cluster of monastic sites at Meteora in Greece. In his analysis, the material-based approach to conservation fails because it creates 'a form of discontinuity... between the monuments and the people, and between the past and the present' (ibid.: 20). The subsequent values-based approach, while attempting to place community at the centre of the conser-vation process, is nevertheless still based on a lack of historical continuity, with the result that 'the aim of conservation remains the preservation of heritage, considered to belong to the past, from the people of the present, for the sake of the future generations (discontinuity)' (ibid.: 22). By contrast, Poulios proposes an alternative framework based on the *continuity* of living heritage with the past, placing what he terms the 'core community' at the centre of the process.

A second and related issue that follows from the above critique is the frequent inability to distinguish between historic monuments, on the one hand, and living buildings on the other. Certainly, this distinction is there in principle, with the familiar differentiation in UK legislation between listed buildings and scheduled monuments, and the acceptance of greater scope for change to the former to ensure they remain in beneficial use. In practice, a presumption against change militates against this, resulting in listed buildings effectively becoming 'monu-mentalised'. Françoise Choay (2001) argues that the 'invention' of the idea of the historic monument in the immediate aftermath of the French Revolution was a necessary response to the confiscation of so much historic property from both the Church and the nobility. Whatever the rights and wrongs of the overthrow of the *Ancien Regime*, Choay's argument implies that, central to the change of status that accompanies a building being declared a historic monument, is its removal, sometimes forcibly, from its original cultural context.

Clearly, some historic structures are overtly and legitimately monumental in this sense, as Alois Riegl (1982) recognised in 1903 in his founding system of conservation values, specifically in the category of 'deliberate commemorative value'. But the potential misattribution of this status to *any* historic building of interest is a political act and one that accounts in large measure for the misunder-standing of conservation by many communities – for example, in the still common assumption that if a building is listed then one cannot change it – making the issue of public participation in conservation immediately contentious. Regrettably, this ambiguity is reinforced by much of conservation theory, which, for example, fails to distinguish between the conservation of artworks such as paintings and sculpture (and, arguably, historic monuments), on the one hand, and historic buildings that are still in use for the purpose for which they were built on the other (see, for example, Muñoz Viñas 2005). The important distinction is that these latter remain situated, in some sense at least, within whatever tradition gave them birth.

Which brings us to a third issue, that of what it is we think we are conserving. Conservation rightly focuses on the material remains of cultures, but in so doing the people who have some sense of ownership of these cultural artefacts are often all but ignored. Defenders of the orthodox methodology will point out that communal value is now one of the four principal categories of values that Historic England's *Conservation Principles* (2008), following the *Burra Charter* (ICOMOS 2000), invites us to identify. However, the weakness of any values analysis is that, having divided a complex whole into more manageable parts, it has little to say about the interconnection of those discrete values; significance is defined, literally, as no more than the sum of its parts. Just as community value has been 'bolted on' to an accretive and unstructured assemblage, so it can all too easily be wholly detached again. It is easy to bracket and gloss over aspects of lesser interest (to whom?) in favour of what the experts deem to be of greater import; such interpretation is, in cultural terms, inevitable, but also political and need not be acknowledged to be effective.

Such analysis into parts, the cornerstone of scientific method, has, of course, proved hugely productive and since the nineteenth century it has recommended itself as the necessary methodology for the credibility of the so-called 'human sciences'. But it comes at a heavy price, since such analysis tells us little about wholes; hence the public caricature of the academic knowing all there is to know about almost nothing at all. The frequent criticism of the scientific method is that it risks destroying the very thing it seeks to investigate; as the Romantic William Wordsworth put it in his poem 'The Tables Turned' (1798),

> Our meddling intellect
> Mis-shapes the beauteous forms of things:–
> We murder to dissect.

This confusion over what we are really conserving is given greater urgency by recent critiques such as that of Laurajane Smith (2006: 44), for whom '"heritage" is not a "thing", it is not a "site", building or other material object.... Rather, heritage is what goes on at these sites.... Heritage… is a cultural process.' In this view, heritage, rightly defined, is primarily intangible, a process and not a product. If we follow this logic, then we are primarily engaged in the conservation of the valued aspects of communities, of which historic buildings are one expression. In a development of these ideas, Waterton *et al.* (2006: 341) note the failure of values-based approaches such as the *Burra Charter* to 'incorporate changing attitudes to community inclusion, participation and consultation' and, applying a form of critical discourse analysis to its text, illustrate its apparently 'uncritical acceptance of a dominant or authorised approach to heritage'. There is much to applaud in this approach, not least the engagement with the political aspect of heritage, but it should be noted that, while acknowledging the importance of the materiality of much heritage, this analysis offers little account of how tangible heritage relates to the intangible; having successfully widened the definition of heritage (and thereby asking much more interesting questions of conservation practice), we are little closer to an understanding of how one cares well for the material aspects of heritage, which remain stubbornly tangible.

Finally, we come to the central question of public participation in heritage. There has always been a popular element to conservation, and indeed much was achieved in the later twentieth century in conservation by mobilising communities to protect their local areas, such as at Black Road in Macclesfield and Taransay Street in Glasgow (Glendinning 2013: 336–7), to name but two. Being thoroughly of modernity, conservation has always given authority to the expert, in our case the trained conservation professional, which, in part, is what made the above two examples of community architecture so radical, since they were not principally expert-led, but mobilised the political resources of the community instead. This privileging of the expert has, thus far, not been substantially challenged by our values-based methodology. This goes against the grain of our postmodern context, in which the idea of a set of 'official' values having greater authority than 'softer' social values is increasingly countercultural and our system's credibility is rapidly eroding. The implications of a values-based system in an increasingly postmodern age are, therefore, potentially very damaging to conservation as currently conceived and we need to wake up to this quickly; if our culture continues to move in a postmodern direction, then the credibility of expert opinion will continue to diminish, given the distrust that postmodernity has for metanarratives.

If conservation remains wedded to a values-based methodology then the fear is that it will leave itself increasingly open to a catastrophic loss of control over the care of the historic environment. Postmodernity is, after all, an answer to Nietzsche's call for 'the revaluation of all values', making the deference to authority on which conservation still relies increasingly unsustainable. Since values can be made to be whatever we wish them to be, who is to say that the values identified by the careful conservator have any more validity than those of the careless destroyer, whether he be the caricature 'greedy developer' or a latter-day James Wyatt? What is argued here as being urgently needed is an alternative framework which allows for public participation without sacrificing the conservation professional's knowledge and experience.

Tradition

In one way or another, all of the above issues flow from the profound irony at the heart of conservation: while conservation is concerned with the well-being of the material production of 'traditional' cultures, modernity is principally concerned, from its roots in the Enlightenment, with the overthrow of tradition.[1] This accounts for the frequent inability to distinguish between living buildings and historic monuments, makes conservation fundamentally conflicted and, perhaps, helps explain why it appears to be so strange a sub-culture to many of those outside it. That this relation is unexamined is unsurprising, since in any age it is exceptionally difficult to stand outside one's own tradition and critique it. It is nonetheless most important that this relationship between conservation and the workings of healthy traditions is examined, as a more nuanced understanding of tradition offers a variety of resources to help address the emerging issues that increasingly question the validity of conservation, some of which are outlined above. Or, to put it more bluntly, in this analysis the future of conservation as a whole rests on arriving at a more satisfactory relation between conservation's laudable aims and its theoretical grounding.

The philosopher Alasdair MacIntyre provides one of the most far-reaching and considered of the recent critiques of modernity. This critique appears principally in three related volumes: *After Virtue* (1985), *Whose Justice? Which Rationality?* (1988) and *Three Rival Versions of Moral Enquiry* (1990). Of these three books, *After Virtue* provides the initial critique of modernity, while the second is concerned with the means by which one tradition can engage with another. *Three Rival Versions*, which was originally given as the Gifford Lectures in Edinburgh in 1988, takes as its focus three publications from the late 1870s: the Ninth Edition of the *Encyclopaedia Britannica*, Nietzsche's *Genealogy of Morals* and Pope Leo XIII's encyclical *Aeterni Patris*. Crudely, one could say that these texts represent modernity, postmodernity and pre-modernity respectively. MacIntyre (1990: 81) asks whether there is any way in which one rival form of moral enquiry might prevail over the others: 'One possible answer was supplied by Dante: that narrative prevails over its rivals which is able to include its rivals within it, not only to retell their stories as episodes within its story, but to tell the story of the telling of their stories as such episodes.' MacIntyre's argument is that only the Aristotelianism of Thomas Aquinas, represented however imperfectly by the last of his three 'rival versions', is capable of both presenting a coherent account in its own right and, crucially, of providing resources to resolve outstanding issues within the other rival traditions.

William Morris and Philip Webb's *Society for the Protection of Ancient Buildings Manifesto* of 1877 arguably marks the birth of modern conservation, and falls precisely into the same time bracket chosen in MacIntyre's formulation. Morris was, of course, vitriolic in his opposition to one expression of modernity in the form of the restoration of historic churches, but at the same time adopted an approach consistent with modernity's opposition to tradition by denying that historically literate change was any longer possible. Recalling previous centuries, even the seventeenth and eighteenth, 'every change, whatever history it destroyed, left history in the gap, and was alive with the spirit of the deeds done midst its fashioning'. But now, for Morris, architecture 'long decaying' has 'died out' and constructive change to historic buildings is therefore now impossible. It should be noted that Morris talks of those buildings 'the living spirit of which, it cannot be too often repeated, was an inseparable part of that religion and thought, and those past manners', but is clearly of the view that that religion, that thought and those manners are all gone for good. The *Society for the Protection of Ancient Buildings Manifesto*, in Poulios's (2104) terms, represents a profound statement of discontinuity. Many of those responsible for the care of these buildings might beg to differ.

We can say that conservation by its nature is concerned with the objects of tradition(s); it would be easy to say the 'remains of tradition', but this would be to beg the question. It is my contention that the unquestioned acceptance of modernity's (mis-)understanding of tradition accounts for much of the tension that is evident in battles, large and small, over the critical ethical question with which conservation is concerned: that of what we should do to best care for a given historic building. Central, therefore, to this argument is an attempt to open up the question of conservation's understanding of tradition.

One aspect of tradition articulated by MacIntyre (1990) is that it is only in commitment to a tradition that one is able to find one's voice. To the extent that the academy as currently conceived shares modernity's belief in, indeed insistence on, a detached lack of commitment, then to say that one speaks from within a specific tradition renders one an unreliable witness. This stance towards non-commitment is also a factor in encounters between conservation professionals and communities. Modernity's view is that it is only the expert who is able to stand back and accurately appraise a situation, with the community view, seen as inevitably compromised by its local commitments, at best adding a little 'local colour'. By contrast, the view from tradition is that we all have such commitments, and to believe otherwise is to deceive ourselves. This is well understood within the field of hermeneutics – for example, in Gadamer's (2004) discussion of prejudice or pre-judgment. Indeed, part of MacIntyre's critique is that modernity is itself one such tradition, but one that denies itself this status and its attendant resources. To deny tradition in this way is to engage in the fantasy that it is possible to adopt 'the view from nowhere', to borrow a phrase from Thomas Nagel (1986).

Beyond the charge of illegitimacy, another common assumption will be that tradition is necessarily and irredeemably *conservative*. Certainly the words conservation and conservative are (patently) etymologically related, and the rhetoric of political conservatism in the UK often appeals to 'traditional' (that word again) 'values' (ditto). Of course, the use of the word 'conservative' immediately begs the question of what it is we are wishing to conserve, as touched on above. Perhaps a better understanding is that, in so far as it can be said to be 'conservative', a tradition is the means by which a society 'conserves' its central cultural questions (in the sense of maintaining them in vibrant good health) from one generation to the next, and a good argument can therefore be made that an understanding of tradition is more radical (a return to the root) than conservative in this sense.

Heritage studies is a growing academic sub-discipline concerning itself, in part, with the same material remains of tradition as does conservation. The two areas employ different epistemological frameworks, the latter, as we have seen above, typically working more from modernity's assumptions, and the former employing assumptions that would be termed more 'postmodernist', working within the dominant understanding that reality is socially constructed. That the two adjacent sub-disciplines are often able to ignore one another is perhaps symptomatic of the fracturing of academic (and indeed professional) life into isolated fragments or silos, each seemingly increasingly lacking in its understanding of its neighbours. By contrast, a pre-modern understanding of tradition offers resources that may help us to overcome these divisions and provide a means to unite what presently operate as two parallel but unrelated discourses, to their mutual benefit.

In MacIntyre's work, no tradition operates in isolation, but rather in conjunction with societal practices and an individual and community sense of narrative. Horton and Mendus (1994: 11–12) describe the widening circles that characterise MacIntyre's moral theory:

> For MacIntyre, the narrative of an individual's life is to be understood against the background of the wider social context within which that individual finds himself or herself. This wider social context consists of sets of practices which serve to define the virtues, and those practices, in turn, sustain and are situated within a tradition which provides the resources with which the individual may pursue his or her quest for the good. It is traditions which are the repositories of standards of rationality and which are crucial to moral deliberation and action.

MacIntyre emphasises the central role of practices, and the fact that the principal mode of learning what it means to be skilful in any human endeavour is through apprenticeship to a 'craft' (the Greek *techne*), whether that be stone carving, or medicine, architecture or parenting. This focus on practices should appeal to many within conservation, but, at the same time, one must also acknowledge that fundamental to MacIntyre's understanding of tradition is that it is both dynamic and generative. It is for this reason that MacIntyre's pre-modern understanding of what constitutes a well-functioning tradition offers resources that will allow a craft such as conservation to engage positively with each of the four issues outlined above. A possible response to the first of these – the question of change and the production of cultural heritage – is made most explicit in a consideration of narrative, not least since any craft itself has a narrative; as MacIntyre (1990: 65) suggests, 'To share in the rationality of a craft requires sharing in the contingencies of its history, understanding its story as one's own and finding a place for oneself as a character in the enacted dramatic narrative which is that story so far.' So it is to a consideration of narrative that we now turn.

Narrative

To treat a historic building as a form of text is not a new idea. In particular, it is not uncommon for buildings to be compared to chronicles or biographies, so the idea of a building as a narrative joins the queue of competing textual metaphors. There are, however, distinctions to be drawn between each of these literary forms. Biography is a specific form of narrative, and certainly includes a significant element of authorial shaping – one need only compare the variety of approaches adopted by competing accounts of the life of a famous figure to see how significant a role the selection and interpretation of the 'raw material of events' can be. Who decides what is deemed to be worthy of inclusion is an essential issue for any narrative approach, whether biography, history or, in our case, conservation to address and we will return to this shortly.

One significant difference between biography and other forms of narrative is that, whether or not the subject is dead, biographies are most often presented as *completed* narratives; in this, they should be seen as a special form of narrative that is primarily backward facing. 'Building-as-biography' is, therefore, a perfectly serviceable metaphor for, say, the architectural historian; but it does not adequately address the concerns of buildings that can be termed *living*, where conservation wrestles with what is to be done in the present for the benefit of the community whose health is inseparably bound up with that of the building.

Chronicle is a different sub-genre again. Definitions may vary, but a chronicle can loosely be described as a factual account of historical events and one that follows the order in which those events occurred, so that a chronicle can be seen as a sort of 'temporal catalogue'. Like a biography, therefore, a chronicle is also backward facing, but differs from biography in that it is, or purports to be, factual, eschewing the idea that it is created through a process of selection and interpretation. The philosophy of history is much concerned with the distinctions between chronicle and narrative, and the extent to which history can be said to be factual, or conversely that it is inevitably shaped in the telling – that is, the extent to which history is story.

Richard Evans (2001: 25) relates the approach of the British historian G.M. Trevelyan to that of Leopold von Ranke, the 'father' of modern history,

> who had distinguished in his day between the rigorous principles of source-criticism needed for an accurate representation of events in the past, and the intuitive method needed to establish the 'interconnectedness' of these events and penetrate to the 'essence' of an epoch. It was this latter operation, which Ranke conceived of in Romantic and religious terms, and Trevelyan in literary and aesthetic terms, that made the difference, in the view of both of them, between the chronicler and the historian. History, said Trevelyan, was a mixture of the scientific (research), the imaginative or speculative (interpretation) and the literary (presentation).

Geoffrey Roberts (2001: 8, emphasis original) identifies a '*phenomenological turn* in the philosophical debate about history and narrative', which 'shared an existentialist standpoint in which everyday perceptions and experiences of historical reality is a crucial informant of the theorization of narrative'. The potential role of narrative in our understanding of history, and of culture as a whole, is highlighted by the postmodernist historian Frank Ankersmit, who declared that 'historical narrative, like metaphor, is the birthplace of new meaning' (2001: 243).

This interest in narrative reflects a search for a richer understanding of temporality than can be afforded within the framework of modernity. Paul Connerton (1989: 61) notes how Paul de Man considers 'the idea of modernity' as consisting in:

> a desire to wipe out whatever came earlier, in the hope of reaching at last a point that would be called a true present, a point of origin that marks a new departure. This combined interplay of deliberate forgetting with an action that is also a new origin reaches the full power of the idea of modernity.

Where modernity is fundamentally atemporal, seeing time as a series of unrelated instants, a narrative view of time affords the present what we could call a 'broader temporal footprint', incorporating past, present and future. Saint Augustine presents the idea of this 'threefold present' in Book XI of the *Confessions*: 'It might be correct to say that there are three times, a present of past things, a present of present things and a present of future things. Some such different times do exist in the mind, but nowhere else that I can see' (XI, 20, 26). Paul Ricoeur develops this idea, noting the 'structural reciprocity of temporality and narrativity' (1980: 169), ideas which he subsequently developed in the three volume *Time and Narrative* (Ricoeur *et al.* 2010, published from 1984).

Stephen Crites, in his 1971 essay 'The narrative quality of experience', compares narrative to music as 'the other cultural form capable of expressing coherence through time' (1971: 294). Reflecting on the role of storytelling in traditional folk cultures, Crites (ibid.: 295) believes there is:

> more to narrative form than meets the eye... even for a culture as fragmented, sophisticated, and anti-traditional as ours.... Such stories, and the symbolic worlds they project, are not like monuments that men behold, but like dwelling-places. People live in them. Yet even though they are not directly told, even though a culture seems rather to be the telling than the teller of these stories, their form seems to be narrative.

Crites sees stories as 'dwelling-places' in which people live. This is a rich and evocative metaphor, and one whose validity and usefulness are not compromised by its inversion to address our concerns in this volume; that is, that the 'dwelling-places' around which the identities of communities are structured can helpfully be read as narrative.

Implications

The central metaphor of this proposed approach is, therefore, that a historic building is better understood as an ongoing community narrative than as a pile of discrete 'gobbets' of significance, as the current methodology might be characterised. That is to say that a building is meaningful as much because of its cultural whole as for its parts; any methodology worth its name must provide an account of what sort of whole it is that emerges from the parts and by what means.

One principal benefit of this central metaphor is that it makes space for ongoing cultural production. Following the logic of this metaphor, each generation has the opportunity (and indeed responsibility) to write a chapter in that ongoing narrative. But if the current chapter is to represent a coherent addition to what is usually already a multi-faceted story, then a narrative approach demands that we understand the plot to date as well as we possibly can. What a narrative approach emphatically does *not* do, therefore, is to excuse us from a thorough engagement with the past. This not only underscores the value of 'informed conservation' (Clark 2001), but goes beyond information and knowledge of the past to engaging with the tradition(s) that formed the narrative to date. In MacIntyre's terms, this involves learning a 'second first language' (1990: 114). Or, to put it in Gadamer's language, we enter a meaningful conversation. Yet as he reminds us (Gadamer 2004: 383), 'the more genuine a conversation is, the less its conduct lies within the will of either partner. Thus a genuine conversation is never the one that we wanted to conduct.'

So, in practical terms, what are the resources that a narrative approach might bring to bear on conservation's 'unresolvables' identified above? First, and perhaps most obviously, narrative is a widespread and familiar genre, so much so that we rarely consider all of the narratives with which we engage in the course of day-to-day living, whether we be a passive 'consumer' or an active 'author'. The promotion of narrative as a means to engage with cultural questions is not uncontested; for example, in literary criticism, the philosopher Peter Lamarque (2014) offers one critical voice in what he calls narrative's 'opacity'. I will attempt to address such criticisms in future research, but my anecdotal experience in architectural practice remains that a narrative approach enables many more lay people to engage with a historic building than if one restricts the discussion to the dissection of a form of significance assembled from discrete values. It is not uncommon for people's faces to light up with enthusiasm when they grasp the living nature of their building in the context of a narrative approach.

Crucially, a narrative approach provides an account of change to historic buildings and allows us to tell it apart from harm, a distinction so lacking in the conventional methodology – indeed, it can be said that the broader societal point of narrative is as a way for communities and individuals to negotiate change. Narrative provides a means of understanding character formation in an individual's response to the events through which they live and accounts for the continuity of individual identity through the inevitable changes of life. In short, it is the best mechanism we have for relating being to becoming.

Closely linked to this, therefore, is that a narrative approach provides a framework for understanding buildings as developing personalities, rather than as completed biographies. It is a commonplace of conservation practice that buildings are better cared for if they remain in beneficial use – that is, as 'living buildings' – but beyond that assertion one struggles to find any discussion of how this process can be managed in practice; a narrative account provides the resources for this.

A narrative approach also goes with the grain of tradition, traditionally understood. As touched on above, tradition is more dynamic than static, due to its fundamental generativity. It is essential, therefore, that any holistically adequate framework includes a future dimension, as does narrative. Part of knowing what we should do at the present moment involves an acknowledgement that, whatever the chapter we end up writing in this generation, future generations will wish to write their own and that this is an inevitable aspect of what it means for a building to be termed 'living'. If an inter-generational narrative is to be comprehensive and coherent, it is incumbent on the author of each chapter to ensure that plot lines are left open for the future. This, of course, is current good practice, but it is practice that I suggest is unsupported by the current methodologies which, for a discipline concerned with the interpretation of history, remain curiously atemporal.

If conservation is to remain effective in its care of the historic environment, it will need to become far more involved with meaningful public participation, as touched on at the outset. Here narrative's fundamentally communal nature is a significant asset, and it explains why, in my experience of practice, lay audiences 'get' the argument from narrative so readily. Communal narrative is at root democratic and therefore lacks the 'democratic deficit' that afflicts conventional expert-focused processes. This democratic characteristic extends across temporal divides to give a voice to past iterations of a given community; as G.K. Chesterton observed, 'it is obvious that tradition is only democracy extended through time… It is the democracy of the dead' (1908: 30).

This is not democracy understood as a series of individuals expressing their opinion and then aggregating the results into the choice of the majority; rather, as Ricoeur has pointed out, narrative constitutes community (1980: 176). This is as much of a challenge in a postmodern as in a modern cultural situation, since both are predicated on the exaltation of the individual over the communal. In building terms, narrative therefore allows a means to articulate 'my story' – for example, my attachment to the pews in 'my' church because my grandmother sat in them each Sunday. Such a form of narrative at least affords a framework for the engagement of individuals with the historic environment. But, beyond this, a narrative framework provides a structure within which to ask what is 'our story' – that is, the narrative of the whole community. It is not suggested that such discussions will be easy in an age of individualism, but at least narrative, rightly understood, provides a means of framing such a discussion.

Finally, narrative has the great virtue of being readily understood. Whether or not one argues that humanity is 'hard-wired' for narrative, it is enough for the present purpose to observe that narrative works: as the cliché goes, 'everyone loves a good story.' My own interest in narrative stems from discovering, in the course of architectural practice, that talking in narrative terms enables many non-professionals to engage with the historic building they use and for which they are responsible, in a new way and sometimes for the first time. But then narrative is, at the very least, familiar; as Barbara Hardy puts it, 'we dream in narrative, day-dream in narrative, remember, anticipate, hope, despair, believe, doubt, plan, revise, criticise, construct, gossip, learn, hate and love by narrative' (1968: 5).

Conclusion

An important thread of this argument is the observation that since conservation is a product of modernity, and since modernity is, at the very least, conflicted in its relation to tradition, then it should come as no surprise that conservation itself is similarly conflicted. Clearly, this is not the place to resolve our stance with respect to modernity. For our argument, it is enough to observe the antipathy of the Enlightenment to tradition and to justify the assertion that, at a minimum, the project to rescue the physical remnants of that tradition (of whatever age) within the mindset and using the tools offered by modernity is neither transparent nor straightforward.

In my view our conflicted understanding of tradition is central to resolving the most pressing question of how conservation can overcome its democratic deficit to become genuinely participative without risking the destruction of significance in the built heritage. By understanding the limitations and potential abuse of values it is possible to frame a set of requirements for a more successful theoreti-cal framework that could resolve these tensions and serve conservation better in its next phase of development. I suggest that narrative offers great promise as the foundation of an alternative methodology; a narrative understanding transforms buildings from a backdrop to human action to themselves being characters in the dramatic production that is culture.

One significant impact of such a change will be in the role of the expert. Without doubt, we still need the expert's knowledge and experience, and arguably now more than ever. The task, for both conservation professional and community alike, will be to reassess the role that that knowledge and experience, indispensa-ble as they are, should play. As expectations of public participation grow in future, such resources will be deployed less in the assertion of authority and more in enriching and informing the community's story, and this will demand substan-tial cultural change for all.

Note

1 The conservation of modernist architecture is, perhaps, the exception that proves the rule; certainly the treatment of contemporary production in this way is contested including, at times, the architects that produced it. The issue is the legitimacy of defining a given building principally as a work of creative art.

References

Ankersmit, F.D. 2001. 'Six theses on narrativist philosophy of history'. In G. Roberts (ed.), *The History and Narrative Reader*. London: Routledge

Augustine, Saint, 1961. *Confessions.* London: Penguin

Avrami, E. 2009. 'Heritage, values, and sustainability'. In Richmond, A. and Bracker, A.L. (eds), *Conservation: Principles, Dilemmas and Uncomfortable Truths.* Amsterdam: Elsevier/Butterworth-Heinemann, 177–83

Callinicos, A. 1989. *Against Postmodernism: A Marxist Critique.* Cambridge: Polity Press

Chesterton, G.K. 1908. *Orthodoxy.* New York: Dodd, Mead & Co

Choay, F. 2001. *The Invention of the Historic Monument.* Cambridge: Cambridge University Press

Clark, K. 2001. *Informed Conservation: Understanding Historic Buildings and Their Landscapes for Conservation.* London: English Heritage

Clark, K. 2014. 'Values-based heritage management and the Heritage Lottery Fund in the UK'. *APT Bulletin* 45(2/3), 65–71

Connerton, P. 1989. *How Societies Remember.* Cambridge: Cambridge University Press

Crites, S. 1971. 'The narrative quality of experience'. *Journal of the American Academy of Religion*, 291–311

Evans, R.J. 2001. *In Defence of History*, rev. edn. London: Granta

Fredheim, L.H. and Khalaf, M. 2016. 'The significance of values: heritage value typologies re-examined'. *International Journal of Heritage Studies* 22(6), 466–81

Gadamer, H.-G. 2004. *Truth and Method*, 2nd edn. London: Continuum

Glendinning, M. 2013. *The Conservation Movement: A History of Architectural Preservation – Antiquity to Modernity.* London: Routledge

Hardy, B. 1968. 'Towards a poetics of fiction: 3) an approach through narrative'. *NOVEL: A Forum on Fiction* 2(1), 5–14

Harrison, R. 2013. *Heritage: Critical Approaches.* Abingdon and New York: Routledge

Historic England. 2008. *Conservation Principles, Policies and Guidance for the Sustainable Management of the Historic Environment.* London: Historic England

Horton, J. and Mendus, S. 1994. *After MacIntyre: Critical Perspectives on the Work of Alasdair MacIntyre.* Cambridge: Polity

ICOMOS Australian National Committee 2000. *The Burra Charter: The Australia ICOMOS Charter for Places of Cultural Significance 1999; With Associated Guidelines and Code on the Ethics of Co-existence*, rev. edn. Victoria: Australia ICOMOS

Lamarque, P.A. 2014. *The Opacity of Narrative.* London: Rowman & Littlefield International

MacIntyre, A.C. 1985. *After Virtue: A Study in Moral Theory*, 3rd edn. London: Duckworth

MacIntyre, A.C. 1988. *Whose Justice? Which Rationality?* London: Duckworth

Macintyre, A.C. 1990. *Three Rival Versions of Moral Enquiry: Encyclopaedia, Genealogy, and Tradition – Being Gifford Lectures Delivered in the University of Edinburgh in 1988.* London: Duckworth

Morris, W. 1877. *The Society for the Protection of Ancient Buildings Manifesto.* London: SPAB

Muñoz Viñas, S. 2005. *Contemporary Theory of Conservation.* Oxford: Butterworth Heinemann

Nagel, T. 1986. *The View from Nowhere.* New York: Oxford University Press

Nietzsche, F.W. 1968. *The Will to Power.* London: Weidenfeld and Nicolson

Poulios, I. 2014. *The Past in the Present: A Living Heritage Approach – Meteora, Greece.* London: Ubiquity Press

Ricoeur, P. 1980. 'Narrative time'. *Critical Inquiry* 7(1), 169–90

Ricoeur, P., Blamey, K., and Pellauer, D. 2010. *Time and Narrative.* Chicago: University of Chicago Press

Riegl, A. 1982. 'The modern cult of monuments: its character and its origin', trans. from German by K.W. Forster and D. Ghirardo. *Oppositions* 25, 20–51

Roberts, G. (ed.) 2001. *The History and Narrative Reader.* London: Routledge

Schofield, A.J. 2009. 'Being autocentric: towards symmetry in heritage management practices'. In L. Gibson and J.R. Pendlebury (eds), *Valuing Historic Environments.* Farnham: Ashgate, 93–113

Smith, L. 2006. *Uses of Heritage.* London: Routledge

Waterton, E., Smith, L. and Campbell, G. 2006. 'The utility of discourse analysis to heritage studies: the Burra Charter and social inclusion'. *International Journal of Heritage Studies* 12(4), 339–55

Wordsworth, W. and Coleridge, S.T. 1798. *Lyrical Ballads, with a Few Other Poems.* London: J. & A. Arch

5 The language changes but practice stays the same

Does the same have to be true for community conservation?

Keith Emerick

Introduction

For many years, perhaps the past 20 years, the explicit objectives of conservation and heritage practice have included the idea that practice should deliver meaningful participation and capacity building.

There is a long tradition of community participation in archaeology, although it is less clear what involvement in an excavation actually delivers in terms of capacity building or lasting engagement. Community participation in conservation projects – such as the community ownership and management of an historic building at risk – is a relatively new field, although there are notable exceptions. The National Trust has a long history of volunteering, but do we know whether the National Trust volunteers have gone back to their homes, built on those skills and become champions of, or instigators for, conservation and capacity building in their communities? Or is it enough that such volunteering is valuable because of its undeniable impact on individual and personal development? What, exactly, is the legacy of such volunteering schemes?

The Heritage Lottery Fund (HLF) in the UK, as the distributor of the proceeds from the national lottery, is a major funder of community heritage projects, but we might question what type of heritage it produces. We can ask whether it has opened up heritage and conservation to different and competing strands and voices or whether it has reinforced an idea of heritage as something property- and fabric-based.

In spite of all the positive debate about the social value and multi-vocality of cultural heritage, stimulated by the first edition of the *Burra Charter* in 1979 (Australia ICOMOS 1999), has the widespread adoption of a values-based approach to conservation and heritage actually or fundamentally *changed* practice? There have been increases in the numbers of community heritage projects, but community ownership, care, maintenance and management of heritage places is relatively rare. Many of the issues limiting that possible shift in emphasis to the community revolve around the role of expertise, the perceived legitimacy, or otherwise, of community interests and a reluctance to see, understand and use a values-based approach to practice as a means of generating change. Too often a values-based approach is used as a mechanism to underscore the value of old fabric or to sustain a national narrative, rather than a tool to help all parties think creatively about how we want to live in the present and what we hand on to the future. I would argue that, in spite of all that has been written about a values-based approach, the big question underpinning all conservation practice still revolves around understanding how we use the past in the present, and I would suggest that we are no further forward in addressing this fundamental concept. Do we want to preserve, protect and hand on places to the future, or do we want to leave our mark on the present and enact our own interpretations of the past?

The issues

The word 'community' has been used a great deal in the opening section, and this reflects the current direction of thought about heritage and conservation practice – that it has to be about 'community' and have 'public benefit'. But whilst practitioners are happy to change the language, are they as happy to change their practice? In this chapter I look at the implications of community engagement and capacity building in conservation from the perspective of the professional practitioner. I think there are two areas where practice needs to change:

- how practitioners engage with people, and
- how practitioners think about and see places.

There are a number of reasons why I consider these two areas important.

Conservation has a long history and, in the development of conservation, different approaches and their attendant languages have come and gone and sometimes been rediscovered. Usages and terms such as 'restoration', 'reconstruction', 'value', 'significance' 'historic environment' and 'cultural environment' have been redefined for new situations, but none of this has necessarily meant new practice. The risk averse nature of conservation practice continues to act as a brake on change. But it is the two great themes elaborated by Ruskin and Morris, and considered central planks of conservation, that remain deeply ingrained in practice: 'anti-restoration' and 'trusteeship'. To the practitioner 'restoration' has now become acceptable within defined limits and as long as it does not go beyond 'conjecture', although in my experience the public consider restoration to be unproblematic and sensible (Emerick 2013). 'Trusteeship' is the idea that no present actually owns the evidence of the past, it just belongs, unchanged to successive futures and, as a consequence, the present, whichever present it happens to be, is never considered worthy enough to add its own mark to the evidence of the past. But how many generations have to pass before people can make change and what happens if people need to use a historic place now?

The values-based approach to conservation offered the possibility of widening participation and engagement in conservation and heritage – but has it delivered? It may be the case that, instead of talking to people about how the values attached to a place could be explored, changed, interpreted or balanced to lead to its enhancement, the values-based approach has merely become a way for experts to *describe* a place, ensuring that expert values remain unchallenged. Instead of being something that offered the potential for practice to be dynamic, it has arguably remained static. The practical outcome is that, instead of moving towards an idea of conservation that might be about consultation, creativity or renewal, and accepting that the past is continually made in the present (Smith 2006: 2), the fear of 'conjecture' and the certainty of 'trusteeship' are used as a shorthand for resisting change.

Is it going too far to suggest that the heritage and conservation sector has concerned itself for far too long with polishing technical terminology and definitions rather than addressing the content of what is being said, written or practised; that perhaps, the heritage sector cannot see the wood for the trees?

However, times change and new demands emerge. We can consider the pressures now facing the heritage and conservation sectors and whether they make the idea of community conservation and community engagement rather more pressing. Although the heritage and conservation sectors might argue that there have never been enough resources, the economic crash and slow recovery since 2008/9 have brought a number of factors into sharp relief.

First, a lack of resources: money is in very short supply and has become more competitive to receive and its distribution tends to be prioritised based on ideas of 'risk', as defined by the heritage sector. Second is a corresponding lack of resources in personnel: jobs (and expertise) are disappearing from the public sector at a rapid rate. The implication of this is that the infrastructure of people who can advise, assist and support community projects is under increasing strain and 'churn'. Then there is the stress on growth for recovery: this requires speed of decision-making, pragmatism and assertions that conservation does not hinder growth, although we can argue about the degree to which growth is or is not sustainable. The down side is that the need for quick decisions means that long or extended engagement with the public is not always possible, which acts against the demands of community engagement. There are also changing attitudes towards expertise: the advent of social media has made conservation and heritage topics more immediate, with people able to comment on and critique projects and decisions, as well as being able to directly express a desire to become involved in conservation and heritage work.

Another factor is the shortage of heritage craft skills. This is an important but not insurmountable problem, although it is circumscribed by the availability of resources. However, it goes deeper than a shortage of apprentices (which was arguably the case before the economic downturn) and is accompanied by a corresponding shortage in the number of heritage professionals who have the experience to encourage and recognise the characteristics that distinguish good work (see CITB, English Heritage and Historic Scotland 2013).

Finally, a changing demography – an increasing percentage of older people in the population and a more culturally diverse population – is certainly significant in the UK. The age profile has led to some easy but misleading calculations about the numbers of people in retirement who could be volunteers. It is more accurate to say that there is a large number of retired people at the moment 'available' to volunteer. However retirement ages in the future will be extended, meaning that people will have to work longer, and therefore that supposed large pool of retired, able-bodied volunteers will not necessarily exist. Youth is often absent, interestingly, in any conversation about conservation volunteering; the expectation is that the conversation is about the middle aged and retired, but the young are somehow forgotten. There are assumptions here about the 'old' having the same values as the heritage sector, but the young have other values and other ideas of what constitutes heritage.

Faced with these pressures, most in the heritage sector have agreed a consensus that 'volunteering' and the use of volunteers can address many of the issues raised above. Volunteers can be used to monitor places, can carry out research and can initiate projects. Funding streams demand the participation of volunteers at all stages of a project and frequently identify public ownership as the legacy of professional projects – through Trusts or endowments.

However, the relationship between professional and volunteer can generate new questions:

- What if the volunteers have a different way of engaging with and imagining the past in the present?
- What if the volunteers have different attitudes towards the historic fabric, the bricks and mortar, of a heritage place?
- What if volunteers have different ideas about value?
- What if the outcomes of a project are not conservation outcomes?
- How does the professional sector build capacity amongst the public and how can it be sustained?

Basically, the more the heritage sector engages with people, the more it will have to change to accommodate the different views and values of those people if it wishes to remain relevant. Hopefully, the following case studies will address some of these newer and traditional themes and perhaps indicate that the future of community conservation can be positive. The first case study is about how practitioners engage with people, and the second about place.

Case study 1 – people: the workhouse garden, Ripon, North Yorkshire

Ripon is a small market town, officially a city because it has a Cathedral church, on the eastern edge of the Yorkshire Dales in North Yorkshire, England. With a population of about 17,000, Ripon has importance as a tourist centre owing to its race course and close proximity (2km) to the Studley Royal and Fountains Abbey World Heritage Site. Studley's world heritage designation – its Outstanding Universal Value – resides primarily in its categorisation as an example of an early eighteenth-century water garden or 'pleasure garden'. The Cathedral has Anglo-Saxon origins whilst the landscape around Ripon contains important prehistoric and medieval remains. The predominant activity in the area is agriculture, although Ripon is something of a dormitory town for people working in the nearby centres of Harrogate and Leeds.

Located on the north-east edge of the town centre is the former Ripon Workhouse (Figure 5.1), a complex of brick buildings consisting of gatehouse, infirmary, vagrants cells, nursery block, mortuary, Receiving Ward (with 14 cells), Guardians' Room and kitchen garden. Two of the buildings are listed Grade II, meaning that they are 'nationally important and of special interest' and deserving of special consideration in the planning process.

Although a workhouse has stood on the site since 1776, managed by the Justices of the Peace, the current group of buildings was completed in 1855. Just beyond the boundary of the workhouse garden is a large, restored, eighteenth-century villa, formerly owned by the Aislabie family, who granted the land for the original workhouse, but who also owned and developed the landscape and gardens that became the Studley Royal World Heritage Site.

Although the majority of the buildings now have new uses, including a Museum of the Workhouse – and therefore a continuing 'life', one of the most active parts of the complex is the workhouse garden. This is a community and volunteer project, initially funded by the HLF.

The garden is divided into a four-plot rotation system, as in the original work-house garden (Figure 5.2). The original paths have been retained but there the similarities end. A contemporary attitude to sustainability has been employed and although the volunteers have used the site records to identify the historic varieties of fruit and vegetables to grow, they have also decided against using the records

Figure 5.1 The Ripon Workhouse, Ripon, North Yorkshire: the gatehouse © Author

Figure 5.2 The workhouse garden with the Georgian villa behind © Author

as a constraint and have proposed other varieties as a way of starting a conversation with the past if certain varieties are no longer available. For example, although the records mention pears, they do not mention apples. After some discussion, the volunteers decided that this was an error – perhaps apples were so common that they did not require specific mention – so the group have looked at historic varieties of apple common to the north of England and chosen three varieties to grow and, specifically, one taken by the eighteenth-century explorer Captain Cook to stave off scurvy in his voyages across the Pacific. Although the life of Captain Cook does not overlap with the workhouse, this apple variety was chosen for the fact that it was a dramatic and active connection with the past, the story of the north of England and the world – and a world of imagination and exploration. But there was also a very powerful and immediate connection about handling and tending a variety with such a long history. Two varieties of rhubarb are grown – one is Queen Victoria (known to have been grown on the site), but the other, planted next to it, is a variety called Prince Albert (historically the husband of Queen Victoria) – which was not grown there historically. The volunteers just enjoyed the joke of putting the two varieties next to each other in the same bed.

The volume of produce generates a twice-yearly dinner, and, in partnership with other growers, a 'have's' and 'have not's' dinner is held as a way for the group to explore the grim realities of the past. They also supply the local restaurants with fresh, seasonal produce.

Horticulture is one of the most direct ways for people to engage with other people and the past. The activities can be as easy or as demanding as the personnel available and build on existing skills. And a garden always changes. At Ripon the workhouse garden is seen as a means of working with people who are often excluded from conservation and heritage – the unemployed, children and the physically and mentally handicapped, and as such the 'products' of the garden are as much about recovery and rehabilitation as they are about heritage and conservation

Although community-managed gardens are nothing new, in the context of the questions about people and volunteering, they illustrate some of the dilemmas of engagement between heritage practitioners and the public. Quite consciously the volunteers have turned their back on the eighteenth-century building that stands at the top of the site and this is because the language, skills and techniques of building conservation are perceived as more alien and exclusive than those that apply to gardens and horticulture. Buildings are dealt with by 'someone else' and, whilst we as practitioners might want them to become more engaged with the buildings, the garden volunteers find the buildings 'limiting' and restrictive, and the two communities of volunteers do not cross over, even though there could be interesting links to explore between the eighteenth-century house and its owners, the different ideas of patronage, the World Heritage Site and the creation and use of the Pleasure Garden – who physically built it, who 'enjoyed' it and who would have been excluded. The 'conjecture' associated with growing something not identified in the records is seen as something that encourages creativity and discussion; whereas the 'conjecture' associated with the restoration of historic buildings is viewed as limiting and negative. For the volunteers the gardening project is a legitimate way of acting in the present but asking questions about the past, but it is also a project with no end, and this runs counter to the prevailing idea of audit and control, where conservation and heritage projects are measurable and have a beginning, middle and end.

The development of the workhouse garden reveals that the people who manage it have a different attitude towards authenticity. To them, 'integrity' is a more appropriate term because the garden and the activity is about a method of working that is the same now as it was when in use, but also about need, local circumstance, story and change. To a gardener a plant has to be tended and grown, but it can be changed, grafted or hybridised to make something different or a stronger version of the original – and this represents different ways of thinking about and using 'values'. 'Heritage' varieties of plants might have historical value, but using them generates aesthetic and communal value. Making the links and associations between the workhouse garden and Studley Royal could enhance the significance of the World Heritage Site by allowing exploration and enhancement of its social value.

There are numerous other heritage horticultural projects in North Yorkshire (mostly concerning the replanting of historic fruit orchards) and, as they grow in number, the participants talk to and learn from each other, exchanging food and ideas. But there is greater need to talk across the interest groups, for heritage and conservation professionals to see such non-fabric projects as legitimate ways to engage with and use the past in the present and learn what motivates different communities.

Case study 2 – place: St Leonard's Church, Sand Hutton, North Yorkshire

The second case study illustrating an area of change concerns how practitioners think about place. Quite often there is a dramatic polarity between the iconic and the everyday; heritage practitioners are drawn to iconic places and structures, which more often than not are designated as nationally or internationally important. However, even where the condition of the iconic is poor, the solutions for its conservation and use can be limited by over-exaggeration of the importance of its designated status.

'Place' is in fairly common use as a conservation term, entering the language with its current usage through the *Burra Charter* (Australia ICOMOS 1999). It was suggested in the *Burra Charter* that 'place' was preferable to the words 'building', 'site' or 'structure' because it was less value-ridden and encouraged the practitioner (principally) to think beyond the limitations of bricks, mortar and timber, and consider character, setting, context, sound, smell and the wider associations and meanings that go with the place. But how many conservation plans or statements of significance actually follow this advice and think beyond the fabric, or recommend change because it might enhance or creatively change values?

Sand Hutton is a small, largely brick village in the Vale of York about 4km north east of York. Set in an agricultural landscape, the village numbers about 200 people, although the surrounding asparagus fields are worked by a large number of migrant workers, so the population can fluctuate during the year. There is a small village primary school which has learned to accommodate the needs of children who speak languages other than English.

Behind the standing church of St Mary's are the fragmentary ruins of St Leonard's Church (Figure 5.3), the former village church, built around 1150–60 and partly demolished in 1824. The ruins consist of the aisle walls and east end, with only the south wall of the aisle surviving to any great height. St Mary's Church was built to the west of the ruins in the 1840s using stone taken from St Leonard's. People who are married in the new church tend to have their wedding pictures taken in the ruin, which has value as a 'romantic ruin'. The local primary school use part of the cemetery as an outdoor classroom, and they use the ruins for art and writing lessons, so it has also acquired social value.

The ruins are designated as a Scheduled Ancient Monument, meaning that they are nationally important and protected by law. The poor and crumbling condition of the fabric and the lack of a clear solution for its repair meant that it was added to the English Heritage[1] 'at risk' register in 2007. Following a remarkable response from the local community and individual organisations, the village raised £15,000 to put against a grant of £45,000 from English Heritage, enabling the ruins to be consolidated. Preliminary work commenced in 2008, but in that same year a minor earthquake struck Yorkshire and a portion of the south wall was dislodged.

What then followed was a series of discussions about how the repair of the collapsed section was to be undertaken, focusing on the technicalities of

Figure 5.3 The ruins of St Leonard's Church, Sand Hutton, North Yorkshire © Author

restoration versus reconstruction, but framed against awareness of its 'national importance'. Much of what had fallen was core work; however, it was also true that, when thinking about 'the place', the building had been a stone quarry for the new church and had been given patch repairs since the 1840s and, as a consequence, its character was something akin to a garden folly. Although it was a nationally important Scheduled Monument and there was intact archaeological evidence beneath the church and in the lower courses of the walls, the research value of the standing remains was slight. It had greater value as a picturesque ruin with social value, and it was agreed that it should be consolidated as such, which meant that less time had to be given to the niceties of 'conserve as found': the repair had merely to be 'in character'.

However, it was considered that its social value could be enhanced in such a way that its overall significance to the village could be elevated, and could become more than just another consolidated ruin. The opportunity was taken to discuss the ruins with the local primary school to see how they might want to use the ruins and put their own mark on its present and future. It was suggested that the school could hold a competition to design a monument commemorating their village, the repair of the ruined church and the earthquake, with the winning entry made into a church monument and incorporated into the collapsed portion of the ruins. The school took up the challenge and some inspirational art work emerged: children researched other earthquake-hit areas and compared and contrasted where they lived with people and places in Turkey, Pakistan and further afield.

Figure 5.4 The new monument in the nave at St Leonards Source © Author

Instead of selecting a single winner, the children selected their favourite images from a number of entries and combined them in a single design. A stonemason was invited to the school to show how stonemasons worked and how a design on paper could be translated into stone. The stonemason had estimated that he would be in school for two hours, but the children were so fascinated by what he could do he stayed all day, with several children saying they wanted to become stone-masons when they grew up.

In July 2014 the village held a dedication ceremony to mark the completion of the conservation work and the addition of the new monument. Instead of creating something to be inserted into the wall, the monument now has pride of place in the centre of the nave (Figure 5.4).

Conclusion

The values-based approach to conservation and heritage has not resulted in an explosion of community-led and community-managed projects, and arguably it has not produced a different way of 'doing' conservation. Why is this? New ideas about expertise being akin to 'facilitation' and 'enabling' (English Heritage 2000; DCMS 2001), thereby making experts into people who actively engage with communities and encourage projects, have not been developed as they perhaps should have been. It has certainly not led to professionals forsaking the labels of national importance or Outstanding Universal Value in favour of the everyday.

The two case studies have similarities in that they were both staffed by volunteers (in the broadest sense of the word), but they were also about creativity and imaginative ways of thinking about the past, and it is here that, perhaps, the values-based approach has failed – not because of any inherent problem with the approach, but because of the way it is executed. 'Conjecture', as a brake on restoration, and 'imagination' do not sit comfortably together. Inevitably, the risk averse nature of conservation leans towards the certainty of 'trusteeship' and away from change. But the values-based approach should be about change, whether it is the values that are enhanced or changed, or the places themselves.

The temptation, as a practitioner, when presented with activities or questions that seemingly have nothing to do with buildings or monuments, like those about horticulture in the Ripon example, or the introduction of a modern element as in the Sand Hutton example, is to revert to the certainties of trusteeship and be risk averse. But why? I would suggest that the conversation with the past represented by these diverse projects is reward enough because they generate new appreciation of values and significance, but primarily in and between the people who use or live with those places, not the experts who parachute in to start and complete projects. Places that were once gardens or contained buildings, or are dilapidated 'use-less' buildings should be thought of as ripe for renewal. If community engagement is to be taken seriously and meaningfully, practitioners will have to accept the legitimacy of a whole range of ways of engaging with the past and understand that people will want to do things differently and will believe different things. Only in this way will community engagement be sustainable and generate a legacy.

There are perhaps two areas worth consideration in the Sand Hutton case study. The first is that it was absolutely essential to be clear about the character the ruined church had acquired and not be misled by its status. Although it was designated as nationally important, the standing remains were a heavily patched garden folly with its aesthetic and social values in the ascendant. Being clear about this meant that the conservation solution became easier to find and execute, but having done this it was then possible to focus on how the aesthetic and social values could be enhanced whilst leaving the academic or research values (the value of the building as archaeological and historical evidence) untouched. Quite often the designated status of a place or the technical detail of conservation terminology can stand in the way of effecting re-use, so it is essential that practitioners learn to take a step back and really see what they are looking at. Are you looking at the wood or the trees?

Second is the perhaps radical suggestion that when a structure is as ruined and unsound as St Lawrence was, the conservation options should be thought of as limitless rather than the quick reaction approach of 'conserve as found'. Obviously solutions are governed by resources, but this does not preclude solutions based around creativity or imagination. The story of the church as a church has, in some respects, ended, but it is possible to blend its past with how we envisage its present and future, so that the different stories exist in the same place at the same time. All you need to do is ask people what they want, how they would like to use a place or how it might be able to be part of their lives. The school children have built a connection with the ruins, and along with other parts of the village have taken on the ownership of the place. At the moment this connection may be about art or the natural environment, but it has the potential to be more and deeper.

If the consensus about volunteering is right, and I think it is, the heritage sector will have to rethink how it engages with people, certainly if it is to be meaningful and sustainable. Perhaps there could be a different way of thinking about 'trusteeship', but one based on 'trust' between practitioners and the public and the expansion of understanding. When practitioners offer engagement both sides have to know that it is real and have to accept the legitimacy of many different ways of using the past in the present. And both sides have to have their eyes open about places, to accept that places have changed and, like the cells in the human body, are renewed. No one could say that the body is not 'authentic'; it changes, and hopefully acquires understanding as it ages. Maybe, in this way, we can start to see the wood and ignore the trees – and maybe we can change practice.

Acknowledgements

Thanks to Nick Thompson, volunteer Head Gardener at Ripon Workhouse, who commented on the draft text and introduced me to the project.

Note

1 English Heritage was formerly the name for the Historic Buildings and Monuments Commission for England (HBMCE), the organisation that advised government on heritage issues and managed over 400 historic properties. In April 2015, English Heritage was divided split into two sections: Historic England, an advisory and statutory body, and the English Heritage Trust, a charity that manages the national collection of historic properties.

References

Australia ICOMOS 1999. *The Burra Charter*. Canberra: ICOMOS
CITB, English Heritage and Historic Scotland, 2013. *Skills Needs Analysis 2013: Repair, Maintenance and Energy Retrofit of Traditional (pre-1919) Buildings in England and Scotland*. Available at https://content.historicengland.org.uk/content/docs/education/skills-needs-analysis-2013-repair-maintenance-energy-efficiency-retrofit.pdf [accessed 28 August 2015]
DCMS 2001. *The Historic Environment: A Force for Our Future*. London: DCMS
Emerick, K. 2009 'Archaeology and the cultural heritage management 'toolkit'. The example of a community heritage project at Cawood, North Yorkshire'. In E. Waterton and L. Smith (eds), *Taking Archaeology Out of Heritage*. Cambridge: Cambridge Scholars, 91–115
Emerick, K. 2013. 'The rehabilitation of restoration'. In N. Piplani (ed.), *Imagining Conservation: The Next 20 Years*. York: York Conservation Studies Alumni Association

English Heritage, 2000. *The Power of Place.* London: English Heritage

Ruskin, J. 1849. *The Seven Lamps of Architecture.* London: Smith Elder

Smith, L. 2006. *The Uses of Heritage.* Abingdon: Routledge

Waterton, E. 2005. 'Whose sense of place? Reconciling archaeological perspectives with community values: cultural landscapes in England'. *International Journal of Heritage Studies* 11(4): 309–25

Waterton, E. 2008. 'Invisible identities: destroying the heritage of Cawood Castle'. In L. Rakoczy (ed.), *An Archaeology of Destruction.* Newcastle-upon-Tyne: Cambridge Scholars

6 Community involvement in mapping and safeguarding of intangible cultural heritage in India

Nerupama Y. Modwel

Intangible cultural heritage in India

India is home to countless ancient yet constantly evolving cultural practices that continue to inform and influence our present day living. Its customs, innumerable languages, indigenous knowledge systems, richness and depth of music and the arts, both classical and folk, and a fascinating array of craft skills have been passed down orally through generations. The most well preserved of these are either those where members of different ethnic groups appreciate their meaning, relevance and value to their lives or they still earn a living from these activities. This has contributed to a repository of intangible cultural heritage [ICH] elements unique to this part of the world; diverse, given India's geo-cultural existence and historical past, yet representing a composite culture that binds us in a shared heritage (Modwel and Sarkar 2013).

One of the finest examples of this type of inheritance is to be found in the rich weaving traditions of the country, from the intricate *Jamawar* shawl out of Kashmir, the *zari Kancheepuram sari* from Tamil Nadu or *Benarasi* brocade from Varanasi, to the most delicate of *Chanderi* cotton or *Gadwal* handloom from central and south India. Beautiful folk and tribal art forms are seen in the *Pattachitras* of Odisha, *Pithora* paintings of Gujarat and *Kalamkari* work from Andhra Pradesh, and strong storytelling traditions are represented by the *Kavaad* box and *Phad* scrolls of Rajasthan. Numerous dance styles continue to be learned and performed, with sophisticated classical forms like *Bharat Natyam* and *Kathak*, and vibrant folk expressions like *Teyyam* and *Chhau* finding equal benefaction.

Music encompasses age-old classical singing traditions belonging to different *gharanas* and complex musical instruments like the *Sarod*, *Sitar* or the *Rudra Veena*, all requiring years of mastery. Equally well regarded are folk ballads sung to the accompaniment of the ingenious *Kamaicha* or *Banam*. In many parts of India, an itinerant folk singer eulogising local heroes with simple melodies is more often than not continuing an oral ritual going back hundreds of years (DeKultureMusic 2011). Yet, it is an unfortunate reality that the numbers of such vocalists and the makers of such musical instruments are fast depleting.

Figure 6.1 Seraikela Chhau, a unique masked dance, one of three forms of *Chhau* inscribed in UNESCO's Representative List of the Intangible Cultural Heritage of Humanity © INTACH 2012

The transmission of this knowledge has happened over time in subtly different ways, and this is important to understand in the context of community engagement in our mapping and safeguarding attempts. Rituals, social customs and practices have been learned almost without conscious thought by observing and 'doing' by the side of the elders in the family or neighbourhood or village. Their maintenance depends on sharing of information and imbibing knowledge and becoming familiar with it over a period of time during community activities. Local and indigenous knowledge comprising the wisdom and teachings of these communities, and awareness or interpretation of the environment or the universe, is not to be found in written records. The medicinal practices developed, for example, by the *Khasis* in Meghalaya, or the *Kani* ethnic group in Kerala are oral health traditions, unlike the classical health traditions like *Ayurveda* and *Siddha* which are fully classified and codified, having the *Vedas* or other classical texts as their basis (Rajasekharan *et al.* 2015). Oral traditions consisting of folklore, songs and chants are also verbally transmitted, as is other cultural material like oral histories, literature or narratives, and customary laws, which may not be supported by a writing system (Vansina 1985).

However, in the case of crafts or performative traditions like music, singing, dance or dance-drama, there may be a strong *guru–shishya* (teacher–disciple) tradition that ensures their continuation, as in the case of the dramatic *Mudiyettu* ritual theatre of Kerala. In another example, the expertise for *Kalaripayattu*, one of the oldest surviving martial arts, is imparted painstakingly through rigorous instruction started at a very young age for the trainee. It is a skill requiring physical nimbleness, mental discipline and even knowledge of local medical science (Kalaripayattu – one of the oldest martial arts). While its original warfare or self-defence role may have given way to its current expression as an artistic dance form and fitness regime, it stays true to its traditional structure, thus proving the survival capacity of intangible heritage if transferred continually, and with its integrity of form intact, even though its significance for the practitioner may have altered.

Yet there are hundreds of other elements of intangible heritage in India that are disappearing and in danger of escaping our living memory altogether, as more young people get disassociated and detached from their roots and their community's way of life. This is mainly because the feeling of pride in one's home-grown culture and language has become diluted in the face of an easy to embrace monoculture, and the 'hegemony of the big powerful languages' (Sengupta 2009). Moreover, with limited opportunities in the village, there is an understandable exodus by young people to urban areas to earn what is considered a decent and dignified living, which results in further cultural alienation.

Herein also lies the challenge of community engagement, not only in documentation but also in the continued safeguarding and survival of so much of our intangible culture. Creating awareness for it, and bringing people to an understanding of 'safeguarding the logic of continuity' (Department of Arts and Culture, SA 2009) for some of their cultural expressions is an important step in this enterprise. There is also a pressing need to align viable livelihood opportunities with a community's existing crafts and other traditional cultural occupations to bring to it an acceptable level of economic benefit and improvement in its quality of life. Thus, cultural mapping of an ethnic group's unique cultural identity can prove very useful, provided this group appreciates this usefulness and, to an extent, wants to participate in and own the documentation process, and is invested in its outcome and any resulting follow-up activities.

Indian National Trust for Art and Cultural Heritage (INTACH)

The Indian National Trust for Art and Cultural Heritage (INTACH) is a non-profit NGO, dedicated to the conservation and preservation of the architectural, art and material, living and natural, heritage of India. As a pioneer in the protection of India's cultural heritage, it is today the largest institution of its kind in the country. INTACH was founded in 1984 in New Delhi, with a vision to build and grow a membership-based organisation that would spearhead heritage awareness and conservation efforts in India. Since then, the organisation has built a unique network of over 185 chapters, each comprising of local citizens as volunteer members, thus acquiring an inbuilt capacity to work effectively with people and communities in all parts of the country.

Figure 6.2 A Kinnaura woman from Himachal Pradesh is happy to share her tribe's customs and beliefs during cultural mapping © INTACH–TTF/Anubhav Das 2014

The Intangible Cultural Heritage division of INTACH was started in 2008 to explore and enquire into the challenging area of various domains that come under this category. Folklore and memories of the past are valuable and powerful expressions of a people's inheritance: their work spheres, their cultural, social and spiritual values. The aim of the division has been to document and start programmes to safeguard some of the languishing cultural expressions that sustain traditional communities. It has since worked on mapping of indigenous societies, organising an international seminar on endangered languages in India, documenting dying crafts and traditional knowledge, setting up workshops to engage with people, and promoting intangible heritage through a series of festivals, quizzes and other outreach programmes.

Cultural mapping of and with communities

Initiatives to collect and disseminate information on intangible heritage are believed to be extremely important in fostering 'cultural diversity and human creativity' as laid out in UNESCO's *Convention for the Safeguarding of the Intangible Cultural Heritage* (UNESCO 2003). While both of these are crucial elements in enriching society, there is also a need felt to address the economic imperative attached to some of this heritage. Collection and documentation of ICH information by recording it both textually and audio-visually, collecting associated material and researching its current state, are various ways by which we create ICH inventories to disseminate within the community, and to bring before state and local governments, academicians and the general public.

Thus, inventorying has to be followed by efforts to generate dialogue and discussion within communities, across different age groups, genders and socio-economic strata, and with policy makers, on the relevance and compatibility of ancient traditions and wisdom in modern society, encompassing:

- the social relevance of cultural heritage, including resilience of identity and citizenship enabled by pride and participation in one's culture, and strength of social harmony enabled by respect for other cultural traditions;
- the economic relevance of cultural heritage as a means of livelihood and its sustainability through both supply and demand structures, requiring skills

Figure 6.3 Traditional weaving of the Hrangkhol ethnic group of Assam, with potential to be a source of regular income generation © INTACH 2013

generation and indigenous community development, with focus on creative cultural occupations and industries, the promotion of a wider cultural
tapestry within the younger generation, and tapping into synergies between
culture and associated activities like entertainment, heritage tourism, craft
initiatives, architecture, design and so on.

The attempt must be to find, wherever possible, compatibility between traditional
value systems, beliefs and knowledge, on the one hand, and a rapidly evolving,
technology-dependent economy on the other.

At INTACH, we endeavour to go in with deep respect and sensitivity whenever
we take up for documentation a community's intangible heritage: not as experts but
as collaborators. For the success of such projects we need to carry along the aspirations and beliefs of the custodians of living heritage. Along with our local convener,
we aim to set up and involve a local focus group of community members – leaders
or elders, traditional knowledge-holders and young people – to guide, give an understanding of local protocols and to provide a fresh perspective. Throughout the
process, the community's views are incorporated and concerns addressed, with the
objective of arriving at an approximation of what interventions and revitalisation
measures could be required on an urgent basis and which would help in the formulation of further preservation policies. Such recommendations to steer policy, suggesting possible strategies for promotion of the intangible heritage of a community, are
attached to the report at the end, so that it goes beyond a mapping exercise.

We have worked on a cultural mapping methodology with a strong emphasis
on this aspect. This includes an ICH documentation manual and an exhaustive
template that delineates holistic cultural mapping, but more as a guiding tool than
a rigid format, since situations and ground realities can vary from region to
region, even from one village to the next. We sometimes work with other divisions of INTACH, working on architectural, natural or art and material heritage,
in looking at the intangible aspect to physical heritage like monuments, cultural
sites, sacred groves or artefacts, where the physical is fully understood and appreciated once its artistic and spiritual value is gleaned.

A recent example of a joint cultural mapping and resulting conservation
programme is seen in the work done for the erstwhile Portuguese colony in India,
the Union Territories of Diu, Daman, and Dadra and Nagar Haveli. Although
displaying a strong Portuguese influence in its architecture and religious sensi-
tivities, the region is still a mix of different ethnic groups. During the course of
the documentation, diverse communities of the area came forward enthusiasti-
cally to participate and share their views, personal stories and facets of their
culture. As expected, we discovered that there was a heavier loss of intangible
heritage as compared to built heritage. For example, the *Kharwa*, a fishing
community of Diu uses the auspicious *lugda* during wedding celebrations, the
lugda being a bridal sari made by traditional weavers, the *Vanja*. They also use
wedding bangles made by the *Sangharia* people. Both these communities,
however, have only one or two families making these items now, resulting in the
loss of these cultural elements for all three communities, directly or indirectly.

Workshops on ICH documentation

Before starting any cultural mapping, we conduct regional capacity building workshops to acquaint the local people with the mapping process and methodology, since the goal is to let them undertake the documentation from start to finish with guidance by experts whenever required. This is to ensure that we are able to attain some measure of uniformity and standards in the methods of collection. So the participants for these workshops include INTACH's local conveners who are to lead the project, and members of the community who are to be a part of the core research and survey team. The workshop opens with sessions on understanding intangible heritage and the need to pay special attention to this type of heritage which is, by its very nature, vulnerable and, in many cases, in rapid decline. Other sessions cover methodology, camera work and interview techniques, case studies, sharing of field work experiences, IPR-related issues and the ethics of ICH information collection. This approach has resulted in numerous collection projects across India.

The objective of cultural mapping, as explained during the workshop, is to list and document wherever possible all the cultural assets comprising knowledge, material and people that a community can claim as its own. Closely associated with the intangible could be natural heritage that holds special meaning for it, heritage buildings and sites, cultural organisations working in the area and community activity spaces available for its use. As seen in a similar 'cultural resource framework' for mapping, outlined for a Malaysian context by Pillai (2013), we list out cultural elements that will need to be documented in the Indian context.

> *Intangible heritage*: language, traditional knowledge, folklore, oral traditions, customs, rituals, local healing practices, cuisine, fairs and festivals
> *People*: wisdom keepers, shamans, weavers, craftsmen, musicians, singers, actors, dancers, dance-drama artists, painters, building skills workers, potters
> *Traditional cultural occupations*: art and craft initiatives, cottage industries for agri-products, self-help groups, dance or dance-drama troupes, weaving clusters
> *Natural and built heritage*: forest, river, wetlands, sacred grove, *Baoli*, temple tank and other water bodies; local monuments and sites
> *Community spaces and organisations*: village *chaupal*, art and craft centre, museum, government agencies, NGOs, zonal cultural centre, university research groups, local schools

What emerges even in initial discussions with the local people during the workshop is that not all communities are organised into any sort of group or initiative, have access to the above facilities or support from the above organisations, hence underlining the urgency of such programmes. Certain ethnic societies of India, while lacking significant quantities of built heritage, retain a distinctive way of life in close relationship with their land and nature, and express their worldview through creative and innovative forms of living heritage. Unfortunately, this important segment of our society is often also the most isolated, marginalised and out of mainstream culture, geographically and otherwise. It constitutes 8.61 per cent of the total population of India, numbering 104.28 million and covering about 15 per cent of the country's area (Census of India 2011).

The Ministry of Tribal Affairs' (Tribes India n.d.) listing of the tribal people's 'low social, economic and participatory indicators' – as measured by levels of maternal and child mortality, fragmented agricultural holdings and lack of basic amenities like drinking water and electricity – necessitates focusing on these communities as a priority. These markers point to the need to explore traditional avenues for community development and profitable employment, and to promote revenue-generating activities based on locally available resources. This means recognising and respecting the distinct intangible aspects of their culture, providing wider exposure to their arts and crafts, and ensuring transmission of related skills and craftsmanship.

A mapping of the cultural resources that a community possesses can thus have tremendous value. It is an important tool to assess already existing cultural assets, which are also the community's strength, and which can be built upon and tapped for development. It can discover the areas where gaps or shortfalls exist, and where resuscitation will add value.

Over and above the broad framework laid out earlier, the aim of the cultural mapping project is to identify and chart indigenous societies and other groups with respect to their distinct cultural footprint:

- to map cultural practices within the community and understand the correlation of their culture and religious beliefs with their patterns of survival;
- to document the customary laws, economy, bio-cultural diversity of each group, to document gender issues and the role of elders in traditional societies;
- to identify and document aspects that are under threat of being lost and to establish sustainable activities, to identify master practitioners and holders of knowledge;
- to identify and link resources in the area comprising individuals, groups and other government and non-government organisations in collaborations so that combined efforts can be made for long-term conservation, to safeguard the rights of these communities over their land, forest resources, sacred and cultural spaces and knowledge.

The workshop aims at reaching a consensus right away on the importance of documenting the cultural assets of a people. This is a crucial aspect as participants, comprising local INTACH conveners and volunteers, as well as members of the community under study, need to appreciate the benefits to the community of such an exercise. The community members can ensure the correct perspective as facilitator, guide, researcher, or source of information, feel empowered to direct the flow of this self-documentation and can be drivers of the follow-up activities after the completion of the project. The team leader may also bring in experts in local heritage, historians and anthropologists, university researchers and heads of local government agencies for advice, facilitation and help with resources. Initial reservations, clarifications and new ideas are aired in a collective setting so that we learn from each other's experiences. This is followed by modules to discuss the developed methodology, as outlined earlier. Again, examples from different parts of India highlight varied and unexpected scenarios and challenges particular to an area, which may require tweaking of the approach to suit different needs.

The workshop also touches upon the preparatory work required for any project, whether it is creation of an effective team, defining the scope of the project, getting permissions, completing secondary research, making work plans, deciding timelines, budgets and geographic boundaries, or making a list of people to interview after extensive consultations. This is followed by details on the implementation stage of the project, the processing and reporting of the information collected.

The basics of audio-visual recording techniques are explained through a module built around hands-on training. Participants get first-hand experience of collecting ICH and weaving a narrative by breaking up into smaller groups and using the camera, recorder or video camera to interview each other, followed by a review of this exercise. Observations from these workshops show that community members are excited about and interested in being a part of such sessions and being a part of the ensuing self-documentation of their area's heritage.

Figure 6.4 An important segment of the workshop on ICH documentation is brainstorming in a collective setting to understand the 'what' and 'how' of documentation and learn from each other's experience © INTACH 2015

The workshop is highly interactive, with a strong Q&A slant. Participants are encouraged to share vignettes of their local heritage, in particular the most fragile or endangered, and their ideas on what and how they hope to document, given the basic tools. Other crucial topics covered deal with the necessity of background research, significance of community engagement and inclusion, copyright issues related to intangible heritage, requisite recommendations based on the findings, and usefulness of sharing the results with the community at intervals to generate interest and support for the project, thus bringing in newer ideas and participation. Post-project activities, including dissemination of the study to all stakeholders, the 'so-what' of the project, are also deliberated. The participants come to an understanding of the ultimate aim of the mapping, which is to address the concerns of the community, suggest solutions in consultation with it, influence policy if required, promote and help set up initiatives to revitalise languishing traditions, facilitate transmission of skills and explore the potential of ICH to be a vehicle for community development.

ICH is kept alive by its practitioners and the people from whom it originates and, as specified by UNESCO, their inclusion in the process of safeguarding is always stressed. They are best placed to decide the heritage nature and significance of any given expression or practice, hence mapping should not happen without recognition of it by them. If certain communities are lacking the wherewithal to undertake or even suggest mapping on their own, local conveners are encouraged to invite them to participate in a documentation project, and help them record and inventory their intangible heritage.

The segment on interview techniques brings out the open-ended nature of ICH collection, which requires wide-ranging conversations, and not a rigidly followed questionnaire. So, for example, an opening enquiry regarding a song demonstration by a practitioner could lead to eliciting information on the festive occasion on which it is sung and how else that occasion is celebrated; the musical instruments that might accompany the song and who makes these with what materials; the clothes that might be worn especially for the occasion; who sings it generally; who taught him or her this song; whether they are teaching it to somebody else; the crops grown by that community as perhaps mentioned in the song; the types of food made from the grain; the implements used; the flora common to the area; and so on. Rather than professional expertise, what is required is clarity and common purpose in the minds of the interviewer and people of the community on the elements of ICH on which information is being sought.

The workshop sessions also incorporate a site visit to a nearby location to give the participants a chance to contextualise the learning from the different modules in a real setting and ask questions about and record elements of ICH that may exist in that particular setting. For instance, a visit to a complex of mediaeval monuments brings up some of the skills – building crafts, wall painting, stone carving, landscaping – inherent in the making of these. Other intangible elements are seen in the intended use for these – mausoleum, mosque or palace – as well as related stories and legends that built heritage of this kind would have garnered over the centuries.

Figure 6.5 Understanding the significance of recognising and recording the intangible elements inherent in tangible forms of heritage during the capacity building workshop on ICH documentation © INTACH 2013

Another visit, for example, to a sari-making family unit, leads to a lively discussion about the many processes involved: the change from using organic colours to chemicals – and if that was really necessary – the health concerns for the workers; the transmitting of the handmade techniques used and the future of such a small industry uniquely producing an iconic sari. The participants also understand how, in a mapping, it is important to visually document complete processes and get comprehensive information during the course of the interview rather than partial information that may have no real value beyond casual viewing.

Once ideas and concepts are honed, by the end of the workshop, the participants return to their district to prepare and fine-tune their project proposals so that the most viable of these can be implemented. Undertaking such a project involves a commitment of at least a year, and needs constant monitoring and midway appraisal to ensure these stay the course, and the people involved continue to be focused on the end goal. Apart from suggestions at the end of the project report, a dissemination and archiving plan is also worked out with the project leader.

Safeguarding measures

ICH mapping is an initial step. The aim of the mapping is not just to create an inventory, but to align all stakeholders: local communities, especially artists, craftsmen and knowledge-holders among them, schools, universities, researchers, government agencies, NGOs and cultural industries, on what is valued in the local culture, and needs to be recognised and documented at the very least, if not revived. During this effort, elements of ICH which call for planning and protection may come up.

Ultimately, the best preservation will come about through a community-based practice of the ICH in question. So, after the mapping is done, we have to look at ways of creating opportunities for participation of communities in their cultural activities:

- respecting the value of intangible heritage for its people and providing recognition to outstanding practitioners in their respective fields;
- involving them in the local school and college educational programmes, intensifying outreach with workshops and cultural festivals;
- helping them link schemes of employment and income generation with the traditional occupations, bringing the study to the relevant authorities and creative industries;
- helping master practitioners or *ustads* identify and develop training programmes for the young with financial support from local or regional agencies, with perhaps a stipend/subsidy element that incentivises them to broaden their repertoire, and to conduct these training sessions;
- using traditional communication systems and practices to spread information about government schemes and awareness campaigns for literacy, health care, sanitation, agriculture and development;
- advocating a concerted push by the government in providing all possible help to promote and broaden understanding of this heritage, spreading awareness at the local level through the media – for communities, students, government, non-government local bodies and corporate houses;
- creating many more national and local cultural centres for promotional activities like exhibitions and public performances, and museums dedicated to intangible cultural heritage;
- focusing on a key area in ICH work which is to empower communities to document their own heritage and know-how and thereby protect their cultural intellectual property, using the mapping to develop booklets to distribute within the community to make them – especially the young people – aware of their rich culture and instil in them pride for it, and as promotional material for further activities like responsible and sensitive heritage tourism;
- engaging children in an understanding and appreciation of their cultural roots with an inclusion of heritage studies in the school curriculum, and including young adults in the thinking, planning and celebration of our living traditions.

Modwel and Sarker 2013

There is increasing public awareness and dialogue on heritage today, with growing concern for our disappearing traditions. What is needed, though, is a formal system of protection for ICH. Identification and selection of diverse elements of

ICH with the consent and involvement of the custodians of that heritage would lead to the forming of a meaningful national registry. This ongoing exercise would bring out high-value intangible heritage which is on the verge of extinction and needs active government protection, not only at the national level, but also at the level of local government where it would have more teeth and maximum impact, since it would be adapted to serve local needs. Local governments and the community would thus play more of a key role in deciding what is to be classified as heritage of national importance and how it is to be protected (Kakiuchi 2014). This will result in bringing safeguarding and promotional activities to those specific elements of intangible cultural heritage which hold special meaning in a particular region, and which are perceived as such by local communities.

Each state in India has numerous art and dance forms, crafts, social and religious practices and oral traditions, entrusted by one generation to the next. Some have survived for hundreds of years, some are languishing. Our effort should be to acknowledge the significance of our cultural past and its bearing on our present, and to encourage in ourselves and others a desire to save it for the future. The Mizo people's concept of *tlawmngaihna*, or service before self, is not visible but is 'the foundation stone of their socio-behavioural dynamics' (Sawyan 2007). The continued communication of all that we value in this heritage is vital to the preservation of our vibrant culture, and development of society. This involves awareness building and continued advocacy with local communities at the village level, either through elders or headmen, young people, women groups and village *panchayats*. Involving members of the local village, district or state in cultural mapping will ensure their continued engagement on these issues and enable them to pursue safeguarding on their own initiative.

References

Census of India 2011. Available at http://www.censusindia.gov.in [accessed 26 August 2015]

DeKultureMusic 2011. *Babunath Jogi. wmv* 22 August. Available at https://www.youtube.com/watch?v=EdmXzwrnlYY [accessed 26 August 2015]

Department of Arts and Culture SA 2009. *National Policy On South African Living Heritage, First Draft*. Available at http://www.maropeng.co.za/uploads/files/National_Policy_on_South_African_Living_Heritage__ICH_.pdf [accessed 26 August 2015]

Kakiuchi, E. 2014. *Cultural Heritage Protection System in Japan: Current Issues and Prospects for the Future*. Tokyo: National Graduate Institute for Policy Studies. Available at http://www.grips.ac.jp/r-center/wp-content/uploads/14-10.pdf [accessed 26 August 2015]

Kalaripayattu – one of the oldest martial arts. 12 April 2015 http://prd.kerala.gov.in/kalarippayatu.htm

Modwel, N.Y. and Sarkar, S. 2013. *Documenting India's Intangible Cultural Heritage*, New Delhi: INTACH Intangible Cultural Heritage Division

Pillai, J. 2013. *Cultural Mapping: A Guide to Understanding Place, Community and Continuity*. Malaysia: Strategic Information Research and Development Centre

Rajasekharan, S., Vinod Kumar, T.G., Shanavaskhan, A.E., Binu, S. and Pushpangadan, P. 2015. *The Teaching of Indian Traditional Medicine*. Richmond: Botanic Gardens Conservation International. Available at https://www.bgci.org/education/1686/ [accessed 21 August 2015]

Sawyan, P.D. 2007. 'Role of intangible heritage in the development process'. In R. Tandon, *Intangible Heritage: Protecting the Living Heritage of the North East*. New Delhi: INTACH

Sengupta, K. 2009. 'Introduction'. In Sengupta, *Endangered Languages in India*. New Delhi: INTACH

Tribes India, n.d. *Home* [online]. Available at http://tribesindia.com/ [accessed 31 August 2015]

UNESCO 2003. *Convention for the Safeguarding of the Intangible Cultural Heritage*. Available at http://www.unesco.org/culture/ich/en/convention [accessed 26 August 2015]

Vansina, J. 1985. *Oral Tradition as History*. Madison, WI: University of Wisconsin Press

7 An insight into Historic England's approach to community-led conservation

Helen Marrison

Recognising the role and potential of communities in conservation practice is vital: from constructing the values that make a place important, to protecting it for future generations to enjoy. Community-led conservation can help shape identities, distinguish sense of place and retain traditional craft skills. There is a growing body of research which demonstrates how volunteering and participation can contribute towards personal development, wellbeing and happiness. In *Conservation Principles: Policies and Guidance for the Sustainable Management of the Historic Environment*, Historic England sets out its 'approach to making decisions and offering guidance about all aspects of the historic environment, and for reconciling its protection with the economic and social needs and aspirations of the people who live in it' (English Heritage 2008: 13). Emphasis on public participation is prominent in the principles which underpin this approach:

> Principle 1: The historic environment is a shared resource.
> Principle 2: Everyone should be able to participate in sustaining the historic environment.

Public participation is placed alongside the other high-level principles which emphasise the need to understand the significance of places, in order to manage and sustain their values through decisions which are reasonable, transparent and consistent (ibid.). Given that public participation and management of the historic environment are viewed together in the overarching framework for conservation here, this chapter will examine the positive impact that community-led conservation can have for the protection of the historic environment itself, rather than focusing on the benefits for identity and personal development.

It is an interesting and critical juncture at which to take an overview of addressing public participation for Historic England. In April 2015 the organisation formerly known as English Heritage became two separate bodies: the English Heritage Trust, and Historic England. The English Heritage Trust is now a standalone charity which, as custodian of the national collection, manages the 400+ properties open to the public. Historic England, a non-departmental public body, continues to perform statutory responsibilities as the government's advisor on the historic environment. Consequently, without visits to properties and the membership scheme, it is now perhaps less obvious for Historic England how the second Principle of participation can be achieved. This is also especially relevant given the disbanding of the Outreach Team in 2011, who were primarily responsible for the delivery of community-led programmes for (what was then) English Heritage. This chapter will demonstrate how the organisation aims to facilitate community-led conservation and achieve dual outcomes: both public participation and a sustainable future for heritage assets.

Before reporting on the various programmes and projects implemented by Historic England, this chapter reviews what is known about the value of community participation in heritage based on existing public policy research, and what the definition of 'community' might be.

This study should not be viewed as a comprehensive account of Historic England's work in community-led conservation schemes, but an insight into the type of activities it supports. Using case studies throughout to illustrate what is currently being implemented, the chapter concludes with key lessons and recommendations for improving the offer to communities.

Understanding the value of community conservation

The *Taking Part* survey which began in 2006, has heightened awareness of participation in heritage through volunteering as well as visiting (DCMS 2015). *Taking Part* collects data on many aspects of leisure, culture and sport in England, including heritage. Ten years on, the data is a unique resource for understanding participation. In 2005/6 *Taking Part* found that 400,000 people volunteered in heritage, with the most recent research showing that figure has risen to approximately 500,000 (English Heritage 2014a: 5). The historic environment sector has been quick to analyse this public passion for voluntary work, and understand the motivations behind it. Research in 2011 reported that 75 per cent of Heritage Open Days volunteers said that volunteering increased their sense of making a useful contribution (Heritage Open Days 2012: 5), with another study by the Heritage Lottery Fund (HLF) showing their volunteers reporting far higher levels of mental health and wellbeing than for the general population (BOP Consulting 2011).

Historic England, whilst acknowledging the evidence for the individual impacts of heritage volunteering, has also sought to understand how volunteering can help to shape the character of a place. In guidance published in 2014, *Pillars of the Community*, Historic England (2014b: 3) comments that:

> The historic environment plays an integral role in shaping the character of a place. On an everyday level it can make an area more attractive to live in, work in, and visit. On a more emotional level it can provide a community with a route through to its past – a tangible reminder of the lives and experiences of previous generations. It is in renewing this link with the past that makes the role of current communities in the management of their heritage assets so potentially important.

What emerges is a mutually beneficial relationship: by engaging communities in conservation we are able to better manage and protect the historic environment, which in turn gives rise to an opportunity for individuals and communities to develop themselves and strengthen identities.

Recognising that community-led conservation need not be limited or exclusively linked to volunteering, Historic England approaches the idea of 'community' in a number of ways. Each year, it produces the *Heritage Counts* report, a survey of the state of the historic environment. This is a body of research, data and information that is commissioned and collected on behalf of the Historic Environment Forum (HEF).[1] The report has an annual theme which responds to challenges or issues facing the historic environment: in 2006 and 2011 the theme was 'Communities' and 'The Big Society' respectively. In 2006, *Heritage Counts* (English Heritage 2006: 22) acutely summarised the main problem with definition:

> The fact that the word [communities] is plural gives a clue to its meaning. While some may not feel part of any community, many people might identify themselves as belonging to more than one community: the neighbourhood or area in which they live; the community which best expresses their origins in terms of ethnicity, faith or belief or perhaps social class; the community to which they belong in terms of their work or key interests. Community can thus be geographical, cultural and interest-based. Our identity is always complex, with different aspects being evoked in different contexts.

It would be nearly impossible to identify the variety of geographical, cultural or interest-based community groups who may wish to participate in heritage, and this poses a fundamental challenge. Table 7.1 explores a possible structural interpretation of community-led programmes.

Table 7.1 Community action and participation

Level of community action	*Who belongs at the level*	*Description*
Grass roots	Individuals identifying with a community	Geographical, cultural, interest-based
These become... ⬇		
Collective	Community groups	Bring people together for a cause(s): structured and organised with clear outcomes
These work with... ⬇		
Supporting organisations/ funding bodies	National or local organisations whose remit is to support community groups	Can receive applications to enable community organisations to achieve their outcomes, and offer professional advice and expertise

For Historic England, the most effective way to engage is at the 'collective' level, once individuals within a community have identified a uniting cause. For the purposes of this chapter I wish to distinguish community groups into a further three categories:

1. Community *groups* – a collective of like-minded individuals who have rallied behind a heritage asset or cause and been able to demonstrate and active interest in heritage that Historic England has been able to foster. An example of this could be a building preservation trust.
2. Community *representatives* – such as local authorities and civic societies. Where local authorities might not be conventionally viewed as 'community groups', their role in shaping the places in which communities live makes them an essential stakeholder. Other levels of local government can also be considered in this category, such as parish councils. Local authorities and civic societies also fall into the third category.
3. Community project *facilitators* – facilitators include amenity societies or organisations that 'play a vital role in mediating the two-way conversation that needs to take place between public agencies and the diverse communities they serve' (ibid.: 7). These can sometimes be consultancies, or umbrella organisations such as Civic Voice. Facilitators are not necessarily community groups that have formed from the grassroots, but do have a close working relationship with members of the community.

As the government's advisor on the historic environment, Historic England is necessarily limited in what it can achieve directly in facilitating community-led conservation. The role of Historic England is rather as a conduit: creating opportunities and capacity for other groups to lead on the ground work of community participation. In other words: 'As a national organisation, there is a limit to the extent that we can engage ourselves, but of course that is not the point. Instead we want to stimulate greater and better-focused local interest and activity' (Steven Bee, Director of Planning and Development 2003–10, English Heritage 2010: 4).

Stimulating such interest and activity was a core function of English Heritage, and remains so for Historic England. The way in which the organisation supports community-led conservation varies from providing funding and expertise, to policy guidance and celebrating success. In return, communities too are able to offer a supportive role to Historic England and the nation's historic environment. We know that 'different communities are likely to value different elements of the historic environment' (English Heritage 2006: 8) and through identifying what is important to them and helping to conserve it we gain a richer understanding of the nation's story. Community-led conservation is a two-way learning relationship: Historic England learns to better understand and accommodate local heritage values, whilst being able to 'ensure community groups taking on historic buildings have access to the skills and knowledge they need' (English Heritage 2011a: 16).

Historic England and community-led conservation

Using the categorisation of community *groups*, *representatives* and *facilitators* as a framework, the way in which Historic England engages with community-led conservation is clearer. For community representatives, Historic England mainly offers support and advice on the changing planning system, whilst the role in terms of support for facilitators on particular projects is very different. The position of the community actors can be varied also: where we might see volunteers as being responsible for activities such as 'fundraising, organising events, mentoring or coaching, representing or campaigning, engaging in conservation or restoration' (English Heritage 2006: 7), the groups identified here can also act as *observers* of the historic environment.

From 2003–11, the Outreach Team was responsible for 'actively engaging new audiences in participating in, learning from, enjoying and valuing the historic environment'. Each of the English Heritage nine local offices had an Outreach Manager to deliver a range of 'creative, grassroots community heritage projects' (English Heritage, 2011c: 1). Over 500 community projects were undertaken between August 2003 and March 2011, with an estimated 1 million people having participated. One of the key principles of the Outreach Team was not to inspire projects which 'start from a building or a collection to engage people's interest'; instead they were working towards interesting people in heritage in 'its widest sense – the historic environment which is all around them, where they live, work or go to school' (Levin 2010: 42). This approach to community-led conservation, as a legacy of Outreach, remains evident in Historic England's policies; the *Corporate Plan* and *Action Plan* view community participation in heritage as a key outcome in four of their five main objectives (Historic England 2015b). The following account of Historic England's approach to community-led conservation will give an overview of research and guidance; financial and expert support; the role of community groups as observers; the celebration of community-led projects.

Research and guidance

The theme of *Conservation Bulletin 63* in 2010 was entitled *People Engaging with Places*. In it, Steven Bee, then the Director of Planning and Development at English Heritage, comments that: 'If local communities understand and recognise the value of their history, they will expect this to be reflected in the plans and policies put forward by their local authority to sustain that value' (English Heritage 2010: 16).

Local authorities and similar groups are key stakeholders in community-led conservation and Historic England supports this important work with research and guidance on community engagement in the planning process – for example, *Recognising the Historic Environment in Community-led Plans* (Grover Lewis Associates 2010) and *Expanding the Neighbourhood Plan Evidence Base* (Locus Consulting 2014). The former analyses village design statements (VDSs) and parish plans (PPs). VDSs and PPs are guidelines which are set out by local people on the future development of their village or parish. Whilst VDSs and PPs have no statutory role in the planning system (that is they are not involved in planning decision making), they do provide 'a shared vision for the future of the whole community and are about local people contributing their own efforts to bringing about that vision' (Action with Communities in Rural England (ACRE), as cited by Grover Lewis Associates 2010: 5). Understanding how heritage can fit into local planning models is essential to make the case for conservation, and to ensure involvement of the community in that process. Based on this research, Historic England produced *Knowing Your Place* (2011b) – guidance which supports community-led planning in the countryside and 'the incorporation of local heritage within plans that rural communities are producing, reviewing or updating... [in] parish plans and village design statements' in order to:

> show people that the historic features of their locality are not just curiosities from the past, but have a relevance to their own sense of belonging, and to the value of the property they live in. Historic features also reinforce shared community identity and so provide the foundation for what politicians like to call sustainable communities.
>
> English Heritage 2011b: 2

Other recent guidance includes *Pillars of the Community* (English Heritage 2014b) and *Understanding Place* (English Heritage 2012b). These resources help to equip community representatives with the evidence and advice to make local conservation effective by properly integrating community views and perspectives. The 2011 Localism Act reinforced the need for guidance for both local authorities and communities who were being encouraged to take responsibility for community assets – both historical and otherwise. *Pillars of the Community* is designed to provide advice on when and how to transfer heritage assets from public to community ownership, which is especially useful for building preservation trusts[2] and was produced collaboratively with the Architectural Heritage Fund, Heritage Lottery Fund, the Prince's Regeneration Trust, the National Trust and Locality, across the wider sector. In enabling collaborative work in this area, Historic England is able to support community-led conservation using national-level policy tools, such as the suite of guidance for local planning authorities in *Understanding Historic Places* (Historic England 2015c). Here, community representatives can learn about the number of approaches available for identifying the historic dimensions of present day landscapes or townscapes. Capacity building for community-led conservation includes providing elected representatives and local authorities with information: increasing their awareness and appreciation of heritage assets is arguably conservation for communities too.

Support and capacity building

As well as providing guidance to community representatives, Historic England coordinates the Heritage Champion programme. Heritage Champions are local councillors who have been nominated by their authority to represent local views on heritage, influence planning decisions and support historic environment services (such as archaeological and conservation staff). Historic England provides Champions with tools to advocate heritage through publications, training events and a biennial conference. The steady increase in the number of Heritage Champions since the programme began can be measured in terms of percentage of local authorities with a Champion: in 2005/6 the figure was 54 per cent, rising to 74 per cent in 2013/14 (English Heritage, 2014a: 44). The Champions programme has offered a way to help alleviate the negative impact of the reduction in local authority historic environment services: the number of full-time equivalents (FTEs) in this area has reduced by 32 per cent since 2005/6. Heritage Champions are uniquely positioned to represent community values in the planning system, providing a voice for conservation of local heritage in decision making.

Through its National Capacity Building (NCB) programme, Historic England provides funding to build and maintain the capacity of third sector organisations. In 2014/15, this funding amounted to over £1 million and helped 20 national organisations in the heritage sector. Funding to promote community-led conservation at the local level included support officers at the Architectural Heritage Fund, to promote solutions to heritage at risk; and an Industrial Heritage Support Officer based at the Ironbridge Gorge Museum to support the capacity of volunteers working across England to manage historic industrial sites (see Chapter 8). Likewise, a Historic Landscapes Project Officer is employed by the Association of Gardens Trusts (at the time of writing) to support the county gardens trusts in building their capacity to engage in the planning system. This provides essential core staffing to facilitate voluntary sector participation.

NCB funding also contributes towards the operational costs of the national amenity societies, the eight statutory consultees in the planning process (Joint Committee of National Amenity Societies 2010). It enables these organisations to contribute effectively to planning casework, drawing on their specialist expertise and their knowledgeable local voluntary networks, as well as their related activities such as advocacy and community engagement. Amenity societies are key community representatives; their input on heritage planning decisions impacts the places where people live and work, their voice is vital in speaking up for assets of community value. They run successful outreach programmes to broaden community engagement and reach new audiences and Historic England is able to use its funding to help achieve this – for example, in the Young Archaeologists' Club (YAC) run by the Council for British Archaeology. YAC is the only club for young people (aged eight to 16) interested in archaeology, and is made up of a network of clubs across the UK. Involving young adults in conservation at a young age encourages awareness and enjoyment of heritage and involvement in community-led conservation in the future (English Heritage 2014a).

Community-led conservation projects are also supported through Historic England's National Heritage Protection Commissions Team, in line with the priorities of the National Heritage Protection Plan soon to be replaced by the new Heritage 2020 framework. A recent example was the community-based recording of the *London* shipwreck with Wessex Archaeology (Wessex Archaeology 2013).

The role of communities as heritage 'observers'

The potential for communities to act as observers and custodians of heritage is significant and a number of local initiatives have been encouraged by Historic England: local listing, the pilot surveys of Grade II Heritage at Risk buildings and war memorials.

The conservation of the historic environment at a national level is, broadly speaking, centred on assets which meet criteria for designation as nationally significant and warranting special protection. These designated assets must gain additional planning consent for any alterations to be made, a process managed through local planning authorities and – where necessary – in consultation with Historic England and the national amenity societies. Historic England is also keen, however, for communities to identify where assets are important to them locally but may not meet the national criteria. A movement of 'local listing' has emerged, largely coordinated by civic societies and their umbrella body, Civic Voice, which has been supported by Historic England to work with local groups on creating local lists. Whilst local listing has no statutory function, it is an important indicator of community support and enthusiasm for heritage. In 2013/14 just under half of local authorities had a local list and to support growth in this area of local protection, Historic England has produced good practice advice on effective ways to compile a local list. The guidance is designed not to be prescriptive and to encourage communities to develop local heritage lists to best meet their needs (English Heritage 2012a).

One of the significant community-led conservation initiatives led by Historic England in recent times has been the Grade II Heritage at Risk pilot survey project. Heritage at Risk (HAR) began as a way to focus public attention on threatened buildings in their area in the 1980s. What started as buildings at risk (BAR) soon gathered momentum to cover all types of asset. Currently, only Grade I and II* buildings are assessed nationally on the HAR Register, with Grade II buildings included for London only. The purpose of the Grade II HAR pilot survey project was to test different ways in which local groups – communities, societies, local authorities or consultancies – could assess Grade II BAR for themselves. Grade II listed buildings constitute around 93 per cent of the total number of designated buildings in England, and it would be impractical for Historic England to resource a Grade II national survey centrally. Instead, the organisation hopes to be able to harness local enthusiasm for heritage and, in some cases, improve the Grade II BAR registers where local authorities keep them already. The key aims and opportunities were:

- to engage the wider public in making the case for heritage;
- proactive constructive conservation of the historic environment;

- to help find solutions for BAR which are responsible, sustainable and community orientated;
- to offer communities ways to volunteer, thereby improving wellbeing and quality of life.

Nineteen pilot projects were undertaken in 2013 with a range of approaches and models involved. Jura Consultants provided evaluation of the pilots, which has revealed practical insights into Grade II BAR, as well as making recommendations based on the community-led schemes. In total, 4,831 buildings were surveyed, of which 4.2 per cent were found to be at risk and 10.1 per cent were vulnerable (Jura Consultants 2013). The evaluation also established how volunteers were recruited, why they wanted to be involved and some of the outcomes of community involvement. Volunteers were mainly recruited through intermediaries such as civic and amenity societies and universities, suggesting the important role of community facilitators in project delivery. Volunteers were motivated to participate through wanting to protect heritage, gain work experience and raise pride in the local area (ibid.: 56). Crucially, the pilot surveys also showed that, with the correct tools, communities and volunteers are able to provide robust and reliable results on the condition of Grade II buildings. Not only have the Grade II HAR surveys been continued beyond the scope of the pilot project in several instances, but also a second phase of pilot work has just been completed using portable technology such as tablets and smartphones. It is hoped that modernising the programme will make it easier for people to become involved. The HAR pilot schemes are key evidence that community-led conservation, once catalysed, can having a lasting and constructive effect in heritage protection.

Commemoration of historical events can offer a special opportunity to engage communities in conservation. The centenary of World War I has drawn public attention and support and has been the catalyst for several Historic England initiatives. The most notable of these is 'Looking After Our War Memorials', where Historic England is working in partnership with the War Memorials Trust to encourage volunteers to research, record and protect their local war memorial. The project began in 2014 and its outcomes have not yet been analysed, but it will, without doubt, contribute to learning more about how communities can be involved in conservation in the future, adding to the growing evidence base on how best to support and promote such opportunities for local engagement.

Conclusion: raising and recognising pride

Recognition is an important part of community-led conservation both for the people involved, and as a way to inspire and inform projects in the future. Historic England celebrates community-run initiatives – for example, in the annual Heritage Angel Awards, which invite nominees for achievement in all aspects of conservation, from the rescue of a historic building to craftsmanship (Historic England 2015a).

Celebrating the success of projects is just one aspect of encouraging and enabling community-led conservation. There needs to be a clear understanding of who communities are, how to approach them and what information different groups require – be that through producing reports and guidance, or offering individual expertise. Community-led conservation is central to Historic England's conservation principles and the wider sector as a way to achieve mutually beneficial outcomes for communities *and* the historic built environment itself. From the Grade II HAR pilots, in particular, the motivations for volunteering merit special attention, suggesting that community-led conservation projects will be of more value to those involved if they can appreciate significance in the 'bigger picture'. Clear outcomes are therefore crucial to community-led conservation, whether managed by a national or local body. In terms of Historic England's core role, this can be summarised in action to champion community-led conservation through funding, provide experts and supporting expert advice, and offer constructive advice through guidance and research for communities and those that facilitate community programmes.

Ultimately, community-led conservation is able to voice, interpret and protect heritage in a way that national organisations cannot. Without people's enthusiasm, desire to learn new skills and love for the place they live, the condition of England's heritage would be, in its entirety, 'at risk'.

Notes

1 The HEF is a committee of chief executives and policy officers from the main public and non-government heritage bodies in England. Their purpose is to strengthen advocacy work and communications in the sector.
2 Building preservation trusts (BPTs) are not-for-profit organisations which rescue historic buildings for the benefit of the public.
3 Publications issued before April 2015 (see above p. 92) are cited as English Heritage and are now available from the Historic England website.

References

BOP Consulting 2011. *Assessment of the Social Impact of Volunteering in HLF Projects: Year 3*. London: Heritage Lottery Fund

DCMS 2015. *Taking Part* [online]. Available at https://www.gov.uk/government/collections/taking-part [accessed 26 August 2015]

English Heritage[3] 2006. *Heritage Counts 2006: Communities* [online]. Available at http://hc.historicengland.org.uk/archive/Previous-Reports/ [accessed April 2015]

English Heritage 2008. *Conservation Principles: Policies and Guidance for the Sustainable Management of the Historic Environment* [pdf]. English Heritage. Available at https://content.historicengland.org.uk/images-books/publications/conservation-principles-sustainable-management-historic-environment/conservationprinciplespoliciesguidanceapr08web.pdf/ [accessed 26 August 2015]

English Heritage 2010. *Conservation Bulletin 63: People Engaging with Places.* Spring 2010 [online]. English Heritage. Available at https://www.historicengland.org.uk/images-books/publications/conservation-bulletin-63/ [accessed April 2015]

English Heritage 2011a. *Heritage Counts 2011: The Big Society* [online]. English Heritage. Available at http://hc.historicengland.org.uk/archive/Previous-Reports/HC-Economic-Impact/ [accessed April 2015]

English Heritage 2011b. *Knowing Your Place: Heritage and Community-Led Planning in the Countryside*. [online] English Heritage. Available at http://www.historicengland.org.uk/images-books/publications/knowing-your-place/ [accessed April 2015]

English Heritage 2011c. *Outreach: Engaging New Audiences with the Historic Environment* [online]. English Heritage. Available at https://www.historicengland.org.uk/images-books/publications/outreach/ [accessed April 2015]

English Heritage 2012a. *Good Practice Guide for Local Listing* [online]. English Heritage. Available at https://historicengland.org.uk/images-books/publications/good-practice-local-heritage-listing/ [accessed February 2016]

English Heritage 2012b. *Understanding Place: Historic Area Assessment in a Planning and Development Context* [online]. English Heritage. Available at https://historicengland.org.uk/images-books/publications/understanding-place-planning-develop/

English Heritage 2014a. *Heritage Counts 2014: The Value and Impact of Heritage* [online]. English Heritage. Available at http://hc.historicengland.org.uk/national-report/ [accessed April 2015]

English Heritage 2014b. *Pillars of the Community: The Transfer of Local Authority Heritage Assets*. September 2014 [online]. English Heritage. Available at http://www.historicengland.org.uk/images-books/publications/pillars-of-the-community/ [accessed April 2015]

Grover Lewis Associates, 2010. *Recognising the Historic Environment in Community-led Plans: Final Report for English Heritage*. London: English Heritage

Historic England 2012. *Good Practice Guide for Local Heritage Listing* [online]. Historic England. Available at https://www.historicengland.org.uk/images-books/publications/good-practice-local-heritage-listing/ [accessed 26 August 2015]

Historic England 2013. *Underpinning Local Planning Processes NHPP Activity 5B2* [online]. Historic England. Available at http://www.historicengland.org.uk/research/research-results/activities/5b2 [accessed 26 August 2015]

Historic England 2015a. *Angel Awards* [online]. Historic England. Available at http://www.historicengland.org.uk/news-and-features/angel-awards [accessed 26 August 2015]

Historic England 2015b. *Corporate Plan* and *Action Plan* [online]. Available at http://www.historicengland.org.uk/about/what-we-do/corporate-strategy/ [ccessed 26 August 2015]

Historic England 2015c. *Understanding Historic Places* [online]. Available at http://www.historicengland.org.uk/advice/planning/understanding-historic-places [accessed 26 August 2015]

Historic England 2015d. *Planning system* [online]. Available at http://www.historicengland.org.uk/advice/planning/planning-system/ [accessed 26 August 2015]

Heritage Open Days, 2012. *Heritage Open Days Report 2012.* [pdf] Heritage Open Days. Available at https://www.heritageopendays.org.uk/uploads/_document-library/Report2012_Final.pdf [accessed April 2015]

Joint Committee of National Amenity Societies 2010. Members [online]. Available at http://www.jcnas.org.uk/ [accessed 31 August 2015]

Jura Consultants, and Simpson and Brown 2013. Evaluating Heritage at Risk: Reviews of Pilot Projects for English Heritage. London: English Heritage

Jura Consultants 2014. Evaluating Heritage at Risk: Legacy of Grade II Survey Pilot Projects, on behalf of Historic England. London: Historic England

Levin, M. 2010. 'Engaging communities with heritage'. In Historic England, *Conservation Bulletin 63: People Engaging with Places*. Spring 2010 [online]. Historic England. Available at https://www.historicengland.org.uk/images-books/publications/conservation-bulletin-63/ [accessed April 2015], 42–4

Locus Consulting 2014. *Expanding the Neighbourhood Plan Evidence Base: Museums, Records Offices, Archives and HERs* [online]. Historic England. Available at http://www.historicengland.org.uk/images-books/publications/expanding-the-neighbourhood-plan-evidence-base-project/ [accessed April 2015]

Wessex Archaeology 2013. 'Supporting community-based recording: the *London* recording project'. EH (5A4) 6784. English Heritage.

8 Engaging with industrial heritage in the twenty-first century

A perspective from the work of the Industrial Heritage Support Officer

Ian Bapty

Introduction

It is not an exaggeration to state that Britain's industrial past created the modern world. Everything from the mass-produced technology, which shapes twenty-first-century life, to our very view of ourselves began in Britain in the mines, quarries, furnaces, mills, factories and pump houses of the eighteenth and nineteenth centuries (Palmer *et al.* 2012; Trinder 2013). Today, many former industrial sites and buildings are open to the public. From World Heritage Sites to local attractions, they are a key element of British national identity and heritage, and make a major economic and social contribution to the well-being of modern communities (Cossons 2008; English Heritage 2011). Yet the simple fact is that this remarkable legacy by no means equates to a secure future, even for existing preserved industrial heritage sites.

This chapter explores the contemporary issues surrounding the care of Britain's industrial heritage from two distinct perspectives. First, it examines the background to the very idea of preserving and conserving the industrial past in this country, and looks at the particular set of circumstances which have resulted in a strong element of enthusiast and community-led participation in that process. This is in contrast, for example, to the situation in European contexts such as France and Germany, where the recognition and preservation of former industrial sites has essentially been led by top-down government intervention (Oevermann and Mieg 2015). This historical context is important. On the one hand, at a time when austerity agendas are pushing the care of heritage generally towards 'people-led' management models, it insightfully reveals how the care of industrial sites in Britain is a pioneering example of 'bottom-up' community-based heritage management. On the other hand, that historical trajectory also sets the parameters for many of the practical difficulties and limitations the sector now faces, and therefore provides an understanding which is essential to sustaining the distinctive character of publicly accessible industrial heritage sites which now exist in Britain.

The second part of the chapter builds on that analysis by looking at the work of Historic England's project to resource an Industrial Heritage Support Officer (IHSO), as an ongoing, practical example of how industrial heritage sites – and the range of groups and organisations that run them – can be meaningfully supported in the social, political, cultural and financial environment of the early twenty-first century. While the project is England specific, it does have a broader potential relevance, especially in terms of demonstrating how a relatively small investment – in a single paid officer covering a large area – can nevertheless be effective precisely by sensitively building on the 'self-help' approach which is a longstanding characteristic of the individuals and groups who still care for much of Britain's industrial heritage. In that sense, the view which is presented here is not an argument for large-scale 'top-down' investment in the management of industrial heritage, nor a suggestion that such an approach would actually be desirable (a similar view is expressed in the Castleford case study in Chapter 11). Rather the aim is to explore practical ways to reinvigorate and sustain an existing community and volunteer-based management tradition as a continuing part of the British experience of presenting the industrial past.

Discovering and preserving the industrial past in Britain: the historical context

Great Britain is often said to be the world's first industrial nation (Mathias 1983). Whatever the conflicting historical claims in that respect, it is certainly true that, from the seventeenth century onwards, the combination of significant technological developments with the role of Britain as an emerging colonial, mercantile and entrepreneurial power significantly contributed to what is now known as the Industrial Revolution (Trinder 2013). In Britain, that process involved not only the widespread growth of extractive and manufacturing industries (ranging from coal and metal mining to ironworking, textile production and agriculture), but also major social and economic changes which encompassed the rapid development of new urban centres, massive population expansion and the emergence of transformative new ideologies of place, landscape, identity, politics and class (Hudson 1992).

Perhaps in part precisely because of the fundamental nature of this change – which, over two centuries or more variously generated social conflict, far-reaching political implications and increasing nostalgia among some influential groups for all that appeared to have been lost in the transition – the idea that industry was also in itself creating what we might now call 'heritage' was slow to develop. When, in the late nineteenth and early twentieth centuries, independent conservation organisations such as the National Trust and the Society for the Protection of Ancient Buildings were formed, it was precisely the preservation of Britain's threatened pre-industrial past that was their common objective (Jenkins 1995; Donovan, 2007). That same emphasis very much informed the focus of the subsequent twentieth-century development of government-led protection of historic sites via both state ownership and legal protection (Thurley 2013), and even today the popular British perception of a 'historic place' is typically anchored around country houses, mediaeval castles and Roman and prehistoric remains rather than former industrial sites. So, for example, a recent 'Heritage is Great

Britain' national tourism promotional campaign did not feature a single industrial heritage site (Visit Britain 2012).

Against this background, initial approaches to recognising and preserving elements of Britain's industrial past were driven more by narratives of national progress and the evolution of science and technology than by a recognition of the importance of industry in some broader historic sense. So, for example, innovative machines such as George Stephenson's 'Rocket' railway locomotive were seen as worthy of preservation from the mid-nineteenth century onwards (Bailey and Glithero 2000). However, it was really only in the 1920s and 30s that more integrated and broad-ranging concepts of the value of Britain's industrial past began to tangibly emerge. In part this reflected the changing social and economic context of the inter-war years, not least the impacts of the Great Depression, which dealt a serious blow to British 'boom and bust' industries, in particular (notably including Welsh slate and Cornish tin mining, for example), and made clear the reality that some of these industries and the distinctive landscapes and communities they had produced might not last forever (Barton 1967; Lindsay 1974).

From the beginning, the new awareness of industrial history was driven more by working people, including professional engineers, and what would come to be known as 'enthusiasts' working within independent societies and associations, than it was by central government or traditional opinion-makers (Palmer and Neaverson 1998). A notable pioneer in this respect was the Cornish Engines Preservation Committee (later the Trevithick Society), established in 1935 by a small group of enthusiasts specifically aiming to preserve the surviving whim engine at the Levant Mine, near St Just, Cornwall, at a time when mass closure of tin mines in Cornwall was accompanied by the scrapping of many similar engines (Buchanan 1996). The group managed to raise £300 to buy the derelict engine, complete with its engine house, and although it would not operate again until 1991 (by then under the auspices of the National Trust), the principle of forming a society to save abandoned industrial sites had been established.

The formation of industrial heritage preservation associations gathered pace after the Second World War. Such bodies chimed with post-war ideals of citizenship, on the one hand, and nostalgia for declining industry, on the other (Palmer and Neaverson 1998). They also reflected the simple fact that in a time of post-war austerity and with other perceived priorities, even where heritage was concerned (such as the demolition of many stately homes in the 1940s and 50s), neither central government nor other national agencies were focused on preserving the derelict and obsolete remains of industry (Thurley 2013). Unquestionably, there was also a political dimension to this situation. Mainstream ideas of what constituted historic sites had changed little since the nineteenth century, and the heritage of great houses, for example, fitted more easily with prevailing concepts of British identity among political and cultural elites than declining industrial sites, which also presented many economic and social problems in the present day (Samuel 1994). Even established organisations such as the Newcomen Society sometimes inadvertently reflected a comparable view. When they were approached in 1950 concerning proposals to preserve the historically important 'Old Furnace' at Coalbrookdale, Shropshire, where Abraham Darby had first smelted iron with coke in 1709, they advised that there was nothing of value that could be saved on the site. Fortunately, Allied Ironfounders, who owned the furnace, had the vision to take a more enlightened view and the conserved furnace survives today as the centrepiece of the Ironbridge Gorge Museum's Museum of Iron (Darby 2010).

Often the development of preservation groups was associated with the work of determined and strong-willed individuals. For example, the pioneering rescue of the Talyllyn Railway by a volunteer-based society in 1951 – which demonstrated, against the odds, that it was possible for a preservation group not just to save objects from a virtually defunct Victorian mineral railway in a remote Welsh valley, but actually to operate it economically as a self-funding visitor attraction (Boyd 1988) – is closely associated with L.T.C. (Tom) Rolt (Rolt 1953). Rolt was a trained engineer, pacifist and canal enthusiast who had already been involved in setting up the Inland Waterways Association in 1946 (Mackersey 1985). His canal and railway preservation initiatives reflected an ideology which celebrated the lives and culture of working people (Rolt 1944), and in that sense he was interested in preserving disappearing lifestyles as much as industrial heritage in itself. The legacy of key personnel with a strong independent vision remains at many British industrial heritage sites.

Through the 1960s, 70s and 80s this process for preserving industrial heritage via volunteer-based associations and other independent bodies was, in effect, further sanctioned via government policy. While many traditional historic sites were preserved by the state or other country-wide bodies such as the National Trust, the new industrial heritage was typically placed either in the care of small charitable trusts or in public ownership and management via local councils (Palmer and Neaverson 1988). The story of the preservation of Fakenham Gas Works in Norfolk exemplifies that approach. The introduction of North Sea gas saw the rapid demolition of thousands of coal gas installations around Britain, and by the 1980s the small but intact town gas works at Fakenham, which had closed in 1965, was a rare surviving English example. Because of that national significance, it was initially proposed that it should be preserved as a publicly funded national museum operated by the Science Museum. However, that solution was not adopted and it was placed instead in the care of local enthusiasts via the Norfolk Industrial Archaeology Society and the Norfolk Buildings Preservation Trust. Opened as a museum in 1987, it is now England's last surviving coal gas works, and is entirely reliant on volunteer effort to sustain and maintain it (Fakenham Museum of Gas and Local History 2015).

Sometimes the creation of an independent industrial heritage trust was brokered as part of a bigger development/planning-led initiative. The bold project to preserve and redevelop the derelict industrial landscape of the Ironbridge Gorge in Shropshire – which included the world's first iron bridge and the 'Old Furnace' saved by Allied Ironfounders in the 1950s – was closely linked to the adjacent construction of Dawley (subsequently Telford) New Town, which commenced in 1963 (de Soissons 1991). The idea that the new town might include an open air museum where historic buildings could be relocated had been floated in 1965 and, indeed, was part of a conscious process of giving the new town a historically grounded identity (Reynolds 2015). To take this project forward – on a derelict industrial site at Blists Hill near the town of Madeley – and to develop the care of other historic industrial sites in the Gorge, the Telford Development Corporation oversaw the creation of the Ironbridge Gorge Museum Trust. This came into being in 1967 as an independent volunteer-based body (Beale 2014). Initially, with the continued assistance of the Telford Development Corporation, professional staff were recruited, and the Blists Hill museum was opened in 1973. Additional sites were subsequently acquired and developed as museums in the 1970s, 80s and 90s (Figure 8.1), and the Ironbridge Gorge as a whole was inscribed as a World Heritage Site in 1986 (White and Devlin 2007). The Ironbridge Gorge Museum Trust now has 200 staff, 500 regular volunteers and manages ten museums – including former iron, tile, pottery and tobacco pipe works – as well as caring for a range of other industrial monuments, buildings and archaeological remains (Ironbridge Gorge Museum Trust 2015).

Perhaps more typical were smaller 'bottom-up' initiatives, where enthusiasts or local people banded together to save one particular industrial heritage site. A good example is Churchill Forge near Kidderminster, Worcestershire. This rare surviving, water-powered forge had been operated by the Bache family from the nineteenth century, and manufactured a range of spades and other bladed implements, together with specialist tools for the glass industry in nearby Stourbridge. Finally closed in 1969, when the last male member of the Bache family retired, a small group of local people recognised the significance of the forge as essentially representative of the kind of forge which had spawned the Industrial Revolution in places like Ironbridge in the seventeenth and eighteenth centuries. They combined to form a trust to maintain the site and undertake limited opening to the public. The forge itself has remained in the ownership of the Bache family, and continues to be opened by volunteers on a few weekends a year, with the income sustaining the operation of the trust (Churchill Forge 2015). This management model – limited opening with marginal income generation, ownership, often in the hands of third party (either private or sometimes a local authority), and care of the site reliant on the skills and time of a small but dedicated group of volunteers – remains a mainstay of industrial heritage preservation in England, in particular, and is especially relevant to many smaller industrial heritage sites such as pumping engines, forges, watermills and windmills.

From the 1970s onwards, independent industrial heritage organisations have also been supported by a range of 'umbrella' associations which seek both to provide a common forum for groups affiliated to them, and to act as lobbying and pressure groups for the sector. For example, the Association for Industrial Archaeology (AIA) was founded in 1973 to promote the study of what was then the relatively new discipline of industrial archaeology (Buchanan 2014). It now has a network of affiliated local associations, provides grant aid to support industrial heritage restoration and research and organises an annual conference

Figure 8.1 Coalport China Museum occupies the restored buildings of the former Coalport Pottery, and was opened in 1976 by the fledgling Ironbridge Gorge Museum Trust © Ironbridge Gorge Museum Trust

exploring the industrial heritage of a particular locality and other events (Association for Industrial Archaeology 2015). Other bodies of this kind include the Heritage Railways Association (Heritage Railways Association 2015), the Mills Group of the Society for the Protection of Ancient Buildings (Society for the Protection of Ancient Buildings 2015) and the Association of British Transport and Engineering Museums (Association of British Transport and Engineering Museums 2015), to name just a few.

Managing British industrial heritage in the twenty-first century

On the face of it, this British experience provides an important testing ground for the process of managing industrial heritage through the action of volunteer bodies and associations. Today, in England alone, there are around 650 former industrial sites open to the public (Ayris *et al.* 1998), and of these around 30 per cent are directly owned and managed by independent trusts. A similar percentage, although in private or local authority ownership, are also managed and run with the close involvement of a volunteer-based independent body (Bapty 2013). In a time of Europe-wide austerity and significantly reduced government spending on heritage-related projects, the UK seems – for whatever deeply rooted historical reasons – to be ahead of this trend. There is also a perceived fit here with political concepts such as the 'Big Society' (actively promoted under the policies of the UK Coalition Government 2010–15), and related processes such as 'asset transfer' are now very much on the agenda as a mechanism for the devolved care of the historic environment (Historic England 2015a).

Yet, for all that has been achieved through the extraordinary work of independent industrial heritage bodies and the individuals and enthusiasts that have run them for 50 years or more, the legacy today is a mixed one. Much of Britain's most celebrated industrial heritage, including key sites within World Heritage Sites such as the Ironbridge Gorge, Derwent Valley Mills and the Cornwall and Devon Mining Landscape, would not now exist without their efforts. Research undertaken on behalf of English Heritage in 1998 and 2008 nevertheless reveals that the sector is facing widespread problems. This research specifically looked at the state of publicly accessible industrial heritage sites in England. Both the resulting reports – 'Public access to England's preserved industrial heritage' (Ayris *et al.* 1998) and 'Sustaining England's industrial heritage' (Cossons 2008) – identify issues: declining volunteer retention and recruitment; limited technical skills transfer from an ageing volunteer base; and significant difficulties both in adapting to a changing funding and visitor environment and in achieving modern 'best practice' conservation, management and visitor presentation standards. Such issues are exacerbated by the fact that industrial heritage sites are often in relatively poor areas economically, reflecting the decline of traditional industry in the UK, and typically encompass extensive sites, challenging urban locations, big and complex buildings, the care and operation of complicated and obsolete working machinery and difficult interpretative challenges (Figure 8.2).

Figure 8.2 The preserved buildings of Pleasley Colliery (Nottinghamshire) illustrate the scale and complexity of the management challenge typically presented by industrial heritage sites © Ironbridge Gorge Museum Trust

The Cossons report also emphasised the simple fact that the very independence of some industrial heritage groups and the volunteers who run them, who are often the same volunteers who had set the groups up in the first place, could carry its own problems. Attitudes to Health and Safety and prevailing legislation were sometimes out of step with the required modern standard, and the same was true when it came to recognition of Scheduled Monument and Listed Building protection – and, indeed, to understanding 'best practice' distinctions such as the differences between conservation and restoration of historic machinery. So while it was, and is, certainly true that a high level of volunteer input in the care of industrial heritage sites has built a degree of real community ownership of those sites – and this was also shown by the surprisingly positive public perceptions of industrial heritage demonstrated by a 2011 survey undertaken for English Heritage (BDRC Continental 2011) – it has by no means proved to be a flawless management approach. That is also emphasised by the experience of funding agencies such as the Heritage Lottery Fund, which reports relative underfunding of industrial heritage compared to other heritage sectors (National Audit Office 2007). One reason for that is the variable quality of applications from industrial heritage groups, who lack the professional support usually found in bigger organisations.

Declining public spending in the 2010s has only exacerbated this situation and has significantly threatened the sustainability of many smaller industrial heritage sites in particular. These had often relied, in part, on typically small amounts of annual local authority linked grant aid, which is now disappearing – for example, the pending closure of the small independently run Dales Lead Mining Museum at Earby in Yorkshire is, to an extent, a consequence of the loss of a modest £2,500 annual local authority grant aid payment (Tate 2014). At the present time, these kinds of challenges look likely to increase, especially if an ever greater reliance is now placed on independent bodies to take on the care and management of sites and museums which have previously been run by public agencies.

Providing support: the English 'Industrial Heritage Support Officer' model

The 1998 and 2008 English Heritage reports (Ayris *et al.* 1998; Cossons 2008) both suggested measures to support the industrial heritage sector in England and these were reinforced in other documents such as the 'Strategic vision for the effective stewardship of industrial heritage 2008–2013' jointly produced by the Association for Industrial Archaeology (AIA) and English Heritage (Association for Industrial Archaeology and English Heritage 2007). One key recommendation was that smaller volunteer-based industrial heritage groups, in particular, needed access to a source of focused professional help beyond that which was available through existing agencies such as English Heritage or other volunteer-based umbrella bodies, important though these are.

In 2012 English Heritage (now Historic England) therefore funded the new position of the England-wide Industrial Heritage Support Officer (IHSO). The

IHSO is based with the Ironbridge Gorge Museum Trust, and works in partnership with the AIA and the Association of Independent Museums (AIM). Although there are obvious limitations on what one person can do on a country-wide basis, the IHSO has a 'capacity building' brief which includes providing advice, delivering and developing relevant training, promoting the development of networks and new partnerships, facilitating collaborative funding applications and working with other partners such as the AIA to maximise the value of existing support across the sector.

A basic element of the IHSO role has been to serve as a contact point for advice and information. It is true that there is already a great deal of advisory help already available for industrial heritage bodies, including that provided through agencies such as Historic England, by museum development officers, by existing specialist umbrella organisations (such as AIM and the AIA), by the wider charity sector and by informal peer to peer contact. However, the problem for many industrial heritage groups – often without any paid staff, facing significant day to day operating demands and typically disconnected from professional heritage and museum networks – is to find their way to that help in the first place (Cossons 2008). The project has acted to fill that gap, essentially providing a 'clearing house' service where enquiries and requests for support can either be dealt with directly by the IHSO or passed on to other local, regional or national contacts who can provide relevant information.

A typical example of clearing house support involved helping the Ellenroad Steam Museum in Rochdale, Lancashire, to find funding for essential maintenance to the boiler house chimney. Ellenroad preserves and operates a large horizontal mill engine still located in its original engine house, and is run by a trust which relies entirely on volunteers (Ellenroad Engine House 2015). The chimney had been subject a major repair in the 1990s, but by 2013 additional pointing and stabilisation was required. The project did not easily fit funding sources typically tailored to larger schemes. The IHSO suggested a range of alternative options, including the AIA's Restoration Grant programme, the Arts Council's PRISM Fund, locally based grant-giving trusts that may be able to support the work and the potential for local sponsorship. In the end the project was supported by a local donor, and the work was completed in time for that season's steaming days to go ahead.

While this kind of input is typical and funding-related enquiries are frequent, the project has also helped larger charitable and some private/commercial organisations in similar ways. For example, the IHSO advised the Birmingham Conservation Trust on the conservation and restoration of historic machinery at the Newman Coffin Factory as part of an HLF-funded conservation project, identifying specialist contractors who could help. Another rather different case involved assisting the owners of a privately owned bottle kiln at West Hallam, Derbyshire. Advice was offered on appropriate repair strategies relating to water ingress into the former bottle kiln, now converted for commercial use, and on contacts able to help with specialist advice and experience of dealing with similar issues on restored bottle kilns in Stoke-on-Trent. Over 80 industrial heritage sites and organisations have been assisted to date by the project's advice, support and information, delivered nationwide across 23 counties.

Another significant part of the project has been to support and develop industrial heritage 'self-help' networks, and this has demonstrated how focused support can make a difference to small, independently run industrial heritage organisations. A good example is provided by two linked projects aimed at building more effective cooperative and capacity building contacts between various industrial heritage sites in North West England. Although the North West was one of the major power-houses of the Industrial Revolution, and the world's leading textile manufacturing area from the late eighteenth until to the early twentieth century, public presentation of this remarkable legacy is heavily reliant on a handful of museums mainly run by local authorities and independent volunteer groups.

With funding and support from Museum Development North West, a North West Industrial Heritage Network was established in 2014 (Figure 8.3). Three initial meet-ings were hosted by Quarry Bank Mill (National Trust), Helmshore Mill (Lancashire County Council) and the National Waterways Museum at Ellesmere Port. Each day had a specific theme with a linked training session, looking at Volunteering, Health and Safety and Resilience. However, what proved equally important was the oppor-tunity for groups to get together informally, to look at different sites, share experi-ence and to exchange information and resources. Over 40 people from more than 30 organisations attended the days, with strong feedback that the network has proved extremely effective, and three further events took place in 2015.

A second North West network is specifically drawing together groups who manage and run stationary steam engines in the area; this is being supported with the help of Museum Development North West and Historic England's North West Historic Area team. That network has now had five meetings, and is working to develop a skills training programme that will collectively support standards in operating, maintaining and presenting engines. A projected two-year project plan has been developed to include delivery of ten training modules, supporting groups via specialist advisers, and undertaking the collation of existing historic film footage of engines as a modern information resource. This scheme is currently being developed into a detailed HLF Our Heritage application, and if that is successful it is hoped that the project can also be a blueprint for similar industrial heritage training support initiatives in other areas.

The potential to work more effectively through partnership has been a specific emphasis of the IHSO project. This includes actively integrating its work with other national and regional organisations carrying out related support activities. Examples include linking with Historic England and close working with Heritage at Risk and Historic Places teams in the North West, Midlands and London and the South East, in particular; the Heritage Alliance via participation as a trainer and mentor in the current Heritage Lottery funded Giving to Heritage programme; the Association for Industrial Archaeology as a co-opted member of the AIA Council; the European Route of Industrial Heritage participating in ERIK UK events; and, lastly, the Association of Independent Museums/AIM, including participation in their conferences and events.

Figure 8.3 A recent meeting of the North West Industrial Heritage Network hosted by Stott Park Bobbin Mill, Cumbria
© Ironbridge Gorge Museum Trust

Significantly, the project has also begun to develop entirely new partnerships which offer innovative kinds of sustainable future support for industrial heritage bodies, such as a developing relationship with the Institution of Mechanical Engineers. This international professional engineering body already runs the prestigious Heritage Engineering Awards Scheme (Institution of Mechanical Engineers 2015) and the IHSO is now working with the Institution to support the sector in other ways, including using its professional 'clout' to help with issues ranging from insurance to heritage engineering skills training.

An important aspect of the project has been to deliver training for industrial heritage groups, targeting key areas where such support is not already available. An October 2014 training event, funded with the assistance of English Heritage, looked at the specific issue of transferring publicly owned industrial heritage sites into community ownership (Figure 8.4). Given government cost-cutting agendas and the high percentage of preserved industrial heritage remains and buildings currently managed by local authorities (around 30 per cent of all industrial herit-age sites open to the public, see Bapty 2013), this is potentially a big issue for the sector. Using examples including the Chain Bridge Forge in Lincolnshire, which recently went through an asset transfer process, the training day looked at the processes and pitfalls surrounding of transferring industrial heritage sites into the care of charitable trusts. In part using the learning and peer discussion at this event, a 'toolkit' document is now being developed to assist industrial heritage groups considering entering into such a process.

Other training has been tailored to the needs of particular groups and their local circumstances. For example, a 2013 event at the Heartlands attraction in Cornwall was designed to assist industrial heritage sites and groups in the area with develop-ing new fundraising and marketing strategies to cope with the particular problems they face, including the widespread dispersal of industrial heritage attractions in often isolated locations across this large county. The IHSO project has had a major impact in direct training delivery like this, and through participation in wider partner events, reaching 215 organisations via 25 events. Just as importantly, the project's officer has acted as a public advocate for the industrial heritage commu-nity – a voice for the impressive, but largely unheard body of local volunteer-led heritage groups – in numbers of national events, reviews and consultations.

Conclusions

The British experience of managing industrial heritage reflects a long tradition of volunteer and community action coordinated through the action of associations and preservation trusts. While there is much that is positive in that legacy – indeed, sites and heritage assets of national and international importance would not now exist without this enthusiast-led effort – it also means that many indus-trial heritage sites face particular challenges today. This is especially true given the current budgetary pressures on local authorities, and changing contemporary perceptions of what local government exists to do, and the resultant threats to

Figure 8.4 The October 2014 Asset Transfer for Industrial Heritage Sites training event was funded by English Heritage (now Historic England) and brought together a wide range of industrial heritage groups from all over England
© Ironbridge Gorge Museum Trust

both the future of local authority museums and funding available to voluntary groups from that source.

Against this background, the IHSO project has shown that it is possible to deliver meaningful assistance for the broad range of groups and organisations who currently care for the physical remains of our industrial past. That said, it would be wrong overstate the achievements of a single-officer project operating across all of England, or to claim that the project's work has substantively addressed the core underlying issues, such as organisational resilience and volunteer renewal at the younger end of the age range, to name but two. However, strategies such as setting up local networks of industrial heritage sites, and creating targeted training opportunities tailored to identified local needs, clearly do have continuing potential to sustain the unique way in which much of the UK's industrial heritage is managed, not least because such approaches directly build on the sector's distinctive dynamic as a 'bottom-up' peer-led movement. Equally importantly, such support mechanisms are relatively simple and cost effective, and can be enhanced by the development of straightforward web and social media-based information exchange.

Perhaps unsurprisingly, but nevertheless importantly, another key pointer from the project is the need for professional bodies and specialist associations of all kinds to work effectively in partnership to support the industrial heritage sector as a whole. While this has always happened to some extent, it is critical at a time of diminishing resources for all, and when the independence of smaller industrial heritage charities and volunteer groups – which, in one sense, is a positive element of the distinctive British 'enthusiasts' tradition – is also now a real barrier to their continued sustainability. Unless it is possible to build links with these smaller bodies and to draw them into partnership with the wider museum, heritage and voluntary sector, it will be difficult for existing management arrangements for many industrial heritage sites to be maintained. In some ways, there is a threat of a double loss here. Not only are locally and nationally significant heritage assets at risk (see, for example, many of the industrial heritage sites currently identified on Historic England's *Heritage at Risk* register, Historic England 2015b), but the established connections with communities which stem from locally based management may also be at risk just at a time when, in other ways, those links are essential in terms of securing future volunteer recruitment and underpinning successful fundraising and local support.

One way forward here is certainly to be more creative in the kinds of partnerships that are developed to support industrial heritage sites. The project's link with the Institution of Mechanical Engineers is just one example of that potential to look beyond traditional boundaries. However, new connections with other sectors – such as the social enterprise, natural environment and health sectors – clearly offer much novel collaborative benefit for industrial heritage sites, including capitalising on common interests such as volunteer recruitment, and creatively facilitating shared input into wider environmental conservation and well-being agendas (see, for example, the lead set by recent archaeological projects engaging with mental health agendas, McMillan 2013). Moreover, there are obvious wider gains to be had here. These include the capability to draw together innovative cross-sector projects which not only have genuine heritage and community outputs, but also pragmatically offer a unique fundraising 'sell' in a highly competitive environment.

For all that has been achieved in terms of the recognition and care of Britain's industrial heritage, it remains the case that politicians, decision-makers and communities are still ambivalent about the values placed on the physical remains of Britain's industrial past. As a consequence, the preservation of former industrial sites is often difficult. Coal mining sites are an obvious case in point. At the time of writing (August 2015), the Snibston Discovery Museum and Park in Leicestershire, which incorporated the preserved Snibston Colliery, has just been controversially closed by Leicestershire County Council on the grounds of cost saving, even despite strong opposition from local people (Martin 2015). The loss of Snibston as a heritage site is especially notable because it also happens to coincide with the effective end of operational deep mining in Britain. Hatfield Main Colliery unexpectedly closed in June 2015 (*Yorkshire Post* 2015), and the two other remaining pits – Kellingley in Yorkshire and Thoresby in Nottinghamshire – are also set to close in 2015 (Gosden 2014). In the current political and financial environment, and notwithstanding the emotional pull of this 'end of an era' moment, it is very unlikely that any physical remains of these last three deep coal mines will be retained, and there is certainly no prospect of their reinvention as publicly accessible heritage sites.

Yet as change of this kind occurs – and there is a wider pressing contemporary issue in the UK around the recognition and creative reuse of industrial heritage on the 'brownfield' sites which are seen by government as prime target sites for new development (HM Treasury 2015) – it is ever more important that the existing resource of preserved and publicly accessible industrial heritage sites is not overlooked. This is not just because many of those sites are also potentially at risk, despite their nominal preserved status, as potentially illustrated by the Snibston case among others; it is also because those places collectively represent an extraordinary legacy of twentieth-century community, volunteer and local authority effort and commitment to the heritage of working people, within which is a powerful basis to build a new contemporary identity for what industrial heritage of all kinds can be for people today. This chapter has attempted to explore both the context of that opportunity and at least the beginnings of some of the practical ways it can be realised. Perhaps the real challenge is for those who care about industrial heritage in the twenty-first century to achieve as much for our time as the pioneering preservationists of the last century did in theirs.

References

Association of British Transport and Engineering Museums 2015. *Aims and Objectives.* Available at http://www.abtem.co.uk/id1.html [accessed 6 July 2015]

Association for Industrial Archaeology and English Heritage 2007. 'Strategic vision for the effective stewardship of industrial heritage 2008–2013'. Unpublished document

Association for Industrial Archaeology 2015. *About Us.* Available at http://industrial-archaeology.org/about-us/ [accessed 6 February 2016]

120   *Ian Bapty*

Ayris, I., Dormer, I. and Swift Research 1998. 'Public access to England's preserved
industrial heritage'. Unpublished report for English Heritage
Bailey, M.R. and Glithero, J.P. 2000. *The Engineering and History of Rocket.* York:
National Railway Museum
Bapty, I. 2013. 'A business plan for the Industrial Heritage Support Officer Project'.
Unpublished document
Barton, D.B. 1967. *A History of Tin Mining and Smelting in Cornwall.* Truro: Bradford
Barton
BDRC Continental 2011. *Industrial Heritage at Risk: Public Attitudes Survey*. London:
English Heritage
Beale, C. 2014. *The Ironbridge Spirit: A History of the Ironbridge Gorge Museum Trust.*
Coalbrookdale: Ironbridge Gorge Museum Trust
Boyd, J.I.C. 1988. *The Talyllyn Railway*. Didcot: Wild Swan
Buchanan, R.A. 1996. 'Landscape with machines'. In D. Morgan and D. Thackray (eds),
The Remains of Distant Times: Archaeology and the National Trust. Woodbridge:
Boydell Trust, 84–91
Buchanan, R.A. 2014. 'The origins and early days of the AIA'. *Industrial Archaeology
News* 169, 2–4
Churchill Forge 2015. *The Story of Churchill Forge.* Available at http://www.churchill-
forge.org.uk/about.html [accessed 6 February 2016]
Cossons, N. 2008. 'Sustaining England's industrial heritage: a future for preserved indus-
trial sites in England'. Unpublished report for English Heritage
Darby, M. 2010. 'Ironworks to Museum: Coalbrookdale 1709–2009'. In P. Belford, M.
Palmer and R. White (eds), *Footprints of Industry: Papers from the 300th anniversary
conference at Coalbrookdale, 3–7 June 2009.* BAR British Series 523. Oxford:
Archaeopress, 3–15
de Soissons, M. 1991. *Telford: The Making of Shropshire's New Town.* Shrewsbury: Wild Swan
Donovan, A. 2007. *William Morris and the Society for the Protection of Ancient Buildings*.
London: Routledge
Ellenroad Engine House 2015. *About Ellenroad Trust and the Steam Museum Society.*
Available at http://www.ellenroad.org.uk/about-us [accessed 6 July 2015]
English Heritage 2011. 'Saving the age of industry'. *Conservation Bulletin* 67
Fakenham Museum of Gas and Local History 2015. *Our History.* Available at http://faken-
hamgasmuseum.com/about-the-museum/ [accessed 6 July 2015]
Gosden, E. 2014. 'Britain to have just one remaining coal pit after UK Coal announces
closures'. *Daily Telegraph.* 2 April 2014. Available at http://www.telegraph.co.uk/
finance/newsbysector/energy/10740600/Britain-to-have-just-one-remaining-coal-pit-
after-UK-Coal-announces-closures.html [accessed 6 July 2015]
Heritage Railways Association 2015. *Heritage Railways Association Members Site Home
Page.* Available at http://www.hra.uk.com/index.php [accessed 6 July 2015]
Historic England. 2015a. *Pillars of the Community: The Transfer of Local Authority
Heritage Assets*, 2nd edn. London: Historic England
Historic England. 2015b. *Heritage at Risk.* Available at https://historicengland.org.uk/
advice/heritage-at-risk/ [accessed 6 July 2015]
HM Treasury 2015. *Fixing the Foundations: Creating a More Prosperous Nation.* London:
HMSO
Hudson, P. 1992. *The Industrial Revolution.* London: Arnold
Ironbridge Gorge Museum Trust 2015. *About Us.* Available at http://www.ironbridge.org.
uk/about-us/ironbridge-gorge-museum-trust/ [accessed 6 July 2015]

Institution of Mechanical Engineers 2015. *Engineering Heritage Awards.* Available at http://www.imeche.org/about-us/scholarships-and-awards/engineering-heritage-awards [accessed 6 July 2015]

Jenkins, J. 1995. 'The roots of the National Trust'. *History Today* 45(1), 3–9

Lindsay, J. 1974. *A History of the North Wales Slate Industry.* London: David and Charles

Mackersey, I. 1985. *Tom Rolt and the Cressy Years.* London: M. and M. Baldwin

Martin, D.J. 2015. 'Snibston Discovery Museum to close on July 31'. *Leicester Mercury.* 30 January 2015. Available at http://www.leicestermercury.co.uk/Snibston-Discovery-Park-close-July-31/story-25950054-detail/story.html [accessed 6 July 2015]

Mathias, P. 1983. *The First Industrial Nation: The Economic History of Britain 1700–1914*, 2nd edn. London: Methuen and Co.

McMillan, J.I. 2013. 'Making a mark on history with the past in mind'. *Mental Health and Social Inclusion* 17(4), 195–201

Museum and Galleries Commission 1994. *Standards in the Museum: Care of Larger and Working Objects.* London: Museum and Galleries Commission

National Audit Office 2007. *Heritage Lottery Fund.* London: National Audit Office

Oevermann, H. and Mieg, H.A. 2015. *Industrial Heritage Sites in Transformation: Clash of Discourses.* New York: Routledge

Palmer, M. and Neaverson, P. 1998. *Industrial Archaeology: Principles and Practice.* London: Routledge

Palmer, M., Nebell, M. and Sissons, M. 2012. *Industrial Archaeology: A Handbook.* York: Council for British Archaeology

Reynolds, G.C.W. 2015. 'Constructing new town identity: myth, heritage and ImagiNation'. *eSharp* 23(1) [no pagination]. Available at http://www.gla.ac.uk/media/media_404376_en.pdf [accessed 6 July 2015]

Rolt, L.T.C. 1944. *Narrow Boat.* London: Eyre and Spottiswoode

Rolt, L.T.C. 1953. *Railway Adventure.* London: Constable

Samuel, R. 1994. *Theatres of Memory.* London: Verso

Society for the Protection of Ancient Buildings 2015. *SPAB Mills Section.* Available at https://www.spab.org.uk/spab-mills-section/ [accessed 6 July 2015]

Tate, L. 2014. 'Appeal to save Earby's Yorkshire Dales Mining Museum'. *Craven Herald and Pioneer.* 14 August. Available at http://www.cravenherald.co.uk/news/11408764.Appeal_to_save_Earby__39_s_Yorkshire_Dales_Mining_Museum_/ [accessed 6 July 2015]

Thurley, S. 2013. *The Men from the Ministry: How Britain Saved Its Heritage.* London: Yale University Press

Trinder, B. 2013. *Britain's Industrial Revolution: The Making of a Manufacturing People, 1700–1870.* Lancaster: Carnegie

Visit Britain 2012. *Heritage is Great Britain: A Guide for International Media.* Available at http://media.visitbritain.com

White, R. and Devlin, H. 2007. 'From basket-case to hanging baskets: regeneration, alienation and heritage in Ironbridge'. In R. White and J. Carman (eds), *World Heritage: Global Challenges, Local Solutions. Proceedings of a Conference at Coalbrookdale, 4–7 May 2006 Hosted by the Ironbridge Institute.* Oxford: Archaeopress, 47–51

Yorkshire Post 2015. '428 out of work as Hatfield Pit closes a year early'. *Yorkshire Post.* 30 June. Available at http://www.yorkshirepost.co.uk/news/main-topics/general-news/428-out-of-work-as-hatfield-pit-closes-a-year-early-1-7333644 [accessed 6 July 2015]

9 Developing evaluation strategies for engagement projects in museum conservation

Danai Koutromanou

Introduction

Important progress has been made in recent research in the domains of public and community archaeology, as well as in public consultation and participation in area conservation and town planning.[1] There is also a substantial and growing body of literature on the relationship between museums and heritage audiences (see Avrami *et al.* 2000, Crooke 2007, Hooper-Greenhill 2008, Macdonald 2002 and 2005, Merriman 1991, Smith 2014 and others) providing a platform for looking more closely at the relationship of the public with museum conservation practice. Even though, during the past two decades or so, there has been a gradual shift towards public engagement stemming from the public value debate in early 2000s and its impact on current thinking and practice in heritage-related professional fields, effective communication remains a challenge for conservators (Brooks 2008, 2013). Arguably, the field of museum conservation has some ground to make up, in terms of active public engagement and involvement in its current policy and practice.

This chapter focuses on aspects of evaluation of engagement projects in museum conservation, drawing on the researcher's own experience of devising instruments and developing strategies to evaluate the impact of three, separate engagement events. More specifically, it examines the impact on two strands of museum visitors' views. Some of the methodological challenges and practical lessons learnt through this process of developing an evaluation approach are also discussed as potentially relevant for conservation practitioners involved in similar projects.

In recent years, public engagement with conservation has increasingly become an objective for museums and heritage organisations and it is often perceived as a part of the conservator's evolving role and integral to good practice. While various forms of engagement are commonly employed, nonetheless there has been little research into their impact. In addition, their intended aims or how these are being achieved often remain unclear, while appropriate strategies for rigorous and objective assessment of their outcomes seem to be lacking, arguably due to public engagement often being undertaken in a simplistic manner aiming to demonstrate the 'success' of the engagement activity. This approach is commonly found in the sector and more often than not lacks a genuinely participative element (Arnstein 1969: 216). Another striking indication is that independent evaluation is rarely a built-in element in the evaluation of museum projects involving public engagement.

Evaluation has been defined as the 'process of thinking back, in a structured way, on what has worked and why, as a project progresses and reaches completion' (Heritage Lottery Fund 2012). It is not straightforward, however, to determine which component of the project it is that this process focuses on. If the aim is broadly defined to be effective public engagement, then impact is arguably a major variable.

Events in this study are evaluated from the perspective of assessing their impact on visitors' views. Hence, it is essential to first look into the motivations for visiting. People may have a number of reasons for visiting a museum. While leisure is a significant aspect, more commonly reported motivations in the literature mainly refer to informal learning, the educational aspect – considering contemporary museums and heritage sites as informal learning environments. It has also been argued that museum visiting is far more than a learning experience for visitors, and it may be understood as a cultural performance in which people either consciously or unconsciously seek to have their views, sense of self and social or cultural belonging reinforced (Smith 2014). This is a particularly challenging proposition if evaluation is considered in connection with the measurement of impact on visitors' views: if people's reason for museum visiting is to reaffirm pre-existing notions, then it is a question whether we can really talk about impact at all? Whether what is captured is actual change, or simply the recording of pre-existing views, it is still worth exploring how visitors respond to engagement events and what thoughts they express about conservation within these contexts. It has been widely recognised that cultural heritage conservation is culturally and socially dependent (Eastop 2006, 2011; Eastop and Brooks 2011; ICOMOS 1994, 2000, 2014; Pye 2009), and thus the views held by audiences and society as a whole play a determining role in what is defined as good practice.

In relation to the terminology employed, it was decided to adopt the term 'view' rather than 'attitude' to describe the discrete strands of visitors' opinions in relation to certain aspects of cultural heritage conservation. The term 'attitude' is used to describe the overall picture gained from the combination of these strands, as manifested by the views expressed. In the literature, attitudes are described as comprising responses in a number of individual constructs, and such 'constructs' would be comparable to the strands identified in the study. Oppenheim describes attitude as 'a state of readiness or predisposition to respond in a certain manner when confronted with certain stimuli'. She also proposes that 'attitudes are reinforced by beliefs (the cognitive component), often attract strong feelings (the emotional component) which may lead to particular behavioural intents (the action-tendency component)' (1992: 175). According to Bennett *et al.*, this definition indicates that information on attitude needs to draw on three components: 1) the cognitive (what you know), 2) the affective (how you feel about what you know) and 3) the 'action-tendency' (what you are likely to do as a result of what you know and how you feel) (2001: 836).

The term 'views', however, is used throughout the analyses to refer to individual opinions specific to a singular aspect of conservation as these were expressed in each case study. For example, knowledge of the history of a particular object that is part of a museum collection draws on the cognitive component. When a participant refers to this object in analogy with a wild animal in the zoo, their expression arguably draws on the affective component. An attitude is the combination of both these components that leads, for example, to participants attaching greater importance to cultural heritage conservation or developing a greater interest in it.

Methodology

To measure changes in views and explore prevalent attitudes, visitor research was carried out, focusing on two distinct strands of views: 1) on museum objects and 2) on conservation practice. Data was collected via questionnaire and interview. A mixed-method approach was selected: quantitative and qualitative, to answer not only *how much*, but also *what* kind of change occurred. The findings are based on the integrated analyses of the data sourced from three case studies: a permanent exhibition at the Ashmolean Museum entitled Conservation Galleries; an adult learning event at the Yorkshire Museum entitled Conservation Workshop: Pieces of the Past; and the interpretation of conservation projects at Knole House, National Trust. The total sample across the three case studies comprises of 271 survey responses and ten interviews.

The impact of the events was measured via post-visit self-evaluation *in situ*, and on the day of the experience. The two sets of questions consisted of two parts each included in the research instruments (see below). The first part was a closed question measuring: a) whether, according to the participant, any change has in fact occurred and b) how confident they are of the occurrence of that change. The second part was an open-ended question, aiming to define either the type of change that occurred (e.g. increase in the perceived value of conservation), or the reason why no change occurred (e.g. drawing the distinction between persistent pre-existing views and the low impact of an event).

Question A: *Did x event make you think differently about the condition in which exhibits are presented?*

☐ *No, not at all*   ☐ *Maybe a little*   ☐ *Yes, definitely*

Why?..

Question B: *Did x event make you think differently about heritage conservation?*

☐ *No, not at all*   ☐ *Maybe a little*   ☐ *Yes, definitely*

Why?..

Quantitative data was analysed with the use of descriptive statistics. Qualitative data was analysed in accordance with Grounded Theory, an inductive approach focusing on generating or 'discovering' theoretical ideas and explanations from the data, rather than specifying them beforehand. This methodological approach has its origins in the field of sociology and it was initially created with the aim of developing explanatory theories of social processes studied in the environments in which they take place (Glaser and Strauss 1967; Strauss 1987; Glaser 1992; Strauss and Corbin 1990; Charmaz and McMullen 2011). More specifically, this approach was selected because of its close relevance to the nature of this study: researching the *experience* and *understanding* of visitors who are being exposed to information about conservation under different conditions, where the process of engagement takes place. Grounded Theory was also considered appropriate as there has been little previous research on this topic and therefore it was not reliable to formulate hypotheses prior to collecting and reviewing data.

Conservators usually work behind the scenes without being concerned with the impact of their work on people's perceptions of heritage material. Even though during recent years there has been much championing of public engagement, visitors still rarely have a chance to witness conservation practice or learn about the challenges conservators deal with. As a result of this scarcity of applied outreach, identification of case studies was challenging. A pragmatic approach was adopted in order to balance the theoretical inception of this research with the current actuality in museum and heritage institutions on a realistic basis. The case studies chosen effectively illustrate multiple aspects of the subject through different applied forms of outreach.

To delineate the range of public engagement formats, broad categories were identified as representing the most usual ways that visitors may encounter conservation. For the majority of cases, a combination of two or more formats is often employed. Nevertheless, for the purposes of this study, it is useful to identify similarities and differences between the most common types of interactions that people have with conservation, to enable a more rigorous and systematic analysis. While this is not intended as an exhaustive list, most of the formats could be generally classified into the following categories:

Observation – Projects that offer people the opportunity to observe conservation work in action.

Web technologies – Websites, social media, blogging platforms, and forums, commonly used to present the progress of conservation projects to diverse online audiences.

Commercial printed material – Posters, booklets, flyers, postcards, non-specialist books (e.g. coffee table books) introducing conservation to interested lay audiences.

Audiovisual material – Information panels, photographs, videos of conservators in action, digital animation of objects in different states (e.g. restored), radio, film, television programmes.

Information dissemination sessions – Public lectures, conferences, festivals, and seminars.

Exhibitions – Exhibitions that refer directly or indirectly to conservation work, including themes such as conservation techniques, materials, processes, conserved objects, and conservation ethics.

Demonstrations – Deliberate demonstrations of conservation work, conservation studios open or visible by visitors, and live video links of conservators in action.

Interaction – Invited interaction between audiences and professional conservators, usually in the form of Q&A. Interaction can also occur with the use of interactive, audiovisual, and/or tactile exhibits.

Inclusive activities – Engaging people in the context of planned activities such as conservation studio tours, open days, object handling sessions, and conservation workshops.

Participation – This involves people being actively involved in conservation work and/or decision-making through educational programmes, volunteering and/or community consultation.

In this study, three events have been evaluated, selected to explore different contexts and interactions between the public and conservation practice. Based on their format, the engagement projects have been classified into the above six main categories of activity characterised by different modes of engagement.

Research designs are often criticised either for failure to employ rigorous scientific methods, or for being too detached from the real world. Also, in the research methodology literature, boundaries between qualitative and quantitative research have been questioned (Cooper *et al.* 2012). It is widely recognised that all research necessarily involves both 'inductive' and 'deductive' modes of inference, and that any difference between quantitative and qualitative approaches in this respect is a matter of degree (ibid.: 6). This study has attempted to balance between the two by employing mixed methods, collecting both qualitative and quantitative data and approaching the subject from various angles.

The case-study approach was adopted as most appropriate: case studies are said to be 'a step to action' as they emerge from an activity and at the same time contribute to it, while their insights can be directly interpreted and put to use (Adelman *et al.* in Bassey 1999: 23). The case study has been described as the study 'of the particularity and complexity of a single case, coming to understand its activity within important circumstances' (Stake in Bassey 1999: 27). The experimental variables in this research are therefore the different contexts where these attempts occur. The people who take part in the survey are considered as a controlled variable. The research instruments can also be considered as a controlled variable, meaning that the set of questions posed to the participants in each case will be based on the same themes in order to generate, as far as possible, comparable data.

It was considered whether data should be gathered using the exact same instruments and methods across all case studies, with benefits in terms of data

Table 9.1 Format of events examined in the study

Event	Format
Conservation Workshop, Yorkshire Museum	Dissemination, planned activity, participation
Conservation at Knole House, National Trust	Exhibition (information about ongoing conservation work)
Conservation Galleries, Ashmolean Museum	Permanent thematic exhibition

comparability. However, this approach would have significantly limited flexibility to adapt questions to each separate case and the survey design might run the risk of being too detached from or even completely irrelevant to each separate case study, so the instrument and methods used were adapted to the nature of each form of engagement in each different environment. At the same time the questions and the data generated correspond in a broader sense to allow comparison of broad thematic categories that emerged from each case study. The results presented below are for two aspects of participants' experience: its impact on their views about the condition in which museum objects are presented and on their views about heritage conservation practice.

Results

Did the visitor experience impact on views about the condition in which objects are presented?

Quantitative analysis in these three case studies shows that the vast majority (89.9 per cent) agreed that the visiting experience has to some extent shaped their views, while most visitors (54.2 per cent) were very confident that their views have indeed been changed. It is evident that, overall, the events examined in this study had a significant impact on the views of visitors in terms of their way of thinking about the condition of museum objects and collections and about the way these are presented.

In the Conservation Galleries, Ashmolean Museum, the vast majority reported change in their way of thinking. While many said that the exhibition definitely made them think differently about the condition of museum exhibits, most were not confident this change was significant, and the rest stated there was no impact at all. The comparatively modest impact could be explained by the fact that this was a static exhibition with no elements of interaction with conservators, or hands-on activities and limited interactive exhibits and, therefore, required more time and effort on the visitors' part to engage with its intended messages. The impact can be still considered as significant, nonetheless, as it has been established that, according to visitors' self-assessment, change has occurred – indeed, to a large extent.

Although this provides one measure for how impactful this event was, it is impossible to determine the various aspects of change that were brought about through a quantitative analysis. To capture the multiple dimensions of impact,

Table 9.2 Change in views on the condition in which museum exhibits are presented

	Ashmolean Museum (%)	*Yorkshire Museum (%)*	*Knole House (%)*	*Average (%)*
Yes, definitely	41.7	66.7	54.2	54.2
Maybe a little	44.8	33.3	28.9	35.7
No, not at all	13.5	0.0	16.9	10.1
Total	100.0	100.0	100.0	100.0

visitors' qualitative accounts were also explored. One of the most striking changes repeatedly reported had to do with the realisation that objects are not necessarily found in the condition they are presented to museum visitors, but may have been treated, and sometimes to the extent of restoration. It was also understood that treatment of heritage material is a complicated process, which does not happen 'automatically' when an object is acquired by a museum, and that preservation state is not a static condition and cannot be taken for granted.

> We think of the objects we see in museums as simply 'being there'. But now I realise the time, expertise and patience needed to get them to their present state. It makes me think what design that artefacts are presented in.
>
> Female, Under 16, International visitor

Many reported that the exhibition made them more conscious of the fact that historical artefacts are a non-renewable resource and also that objects are not secured forever by simply being part of a collection.

> Looking after these things is important because history won't repeat. We've got to look after the things; it's the only history we will get.
>
> Female, Under 16, Local visitor

> If we don't conserve items, they will be destroyed.
>
> Female, Over 64, International visitor

It was frequently mentioned that finding out about conservation was not only revealing but also a cause of surprise due to the fact that they had very limited or no information at all about the process prior to their visit.

> I didn't know at all about the subject. I had never questioned as to if artefacts had been restored.
>
> Female, 16–25, UK visitor

This realisation was accompanied by feelings of deception related with the unknown stages of treatment that objects may have undergone.

> I think I realised there is more smoke and mirrors than I had realised before... I haven't thought about it too much but I suppose all the objects are treated in some way before I see them. It doesn't make me feel great...
>
> Male, 36–50, UK visitor

Several reported that the exhibition made them more aware of the inevitability of decay of materials and their eventual loss. This idea is very close to definitions of conservation articulated in current theoretical frameworks, according to which conservation is understood as the management of change (English

Heritage 2008: 71) rather than an attempt to preserve certain states (Muñoz Viñas 2005: 17).

> And you know, art works are not neutral items which will never change. So even in a very careful way... protected and restored... still it will change over time.
>
> Male, 51–64, International visitor

> We are not having these things forever so the better we take care of it the longer we will have it. I've never thought of it that way.
>
> Female, Under 16, Local visitor

Finding out about conservation and becoming aware of the fact that objects may be treated before they go on display made many visitors consider whether they need to pay more attention, examine more carefully what they are looking at and be sceptical in terms of their understanding and interpretation of artefacts.

> Made me think more about what I was seeing and the possible decision processes that were behind the particular appearance and presentation of a given object.
>
> Female, 16–25, Local visitor

> Not always shows what you think.
>
> Male, 51–64, UK visitor

Some also thought that objects are presented to them in an inappropriate way because the interpretation provided does not include information about conservation interventions and previous states of preservation.

> It made me wonder whether all the information with an exhibit is correct.
>
> Female, 51–64, UK visitor

> Because they are displayed inappropriately.
>
> Female, 36–50, Local visitor

> Made me think about the 'validity' of artefacts in view.
>
> Female, 16-25, International visitor

Some participant reflections on the effect of the exhibition on their way of thinking indicate a potential change, not simply in visitors' views on the subject in question, but also in their attitude – that is, the 'state of readiness or predisposition' (Oppenheim 1992: 175), composed by strands of views, towards the experience of museum visiting in general.

> Will take me more time to think about what happened to objects before they were put on display.
>
> Female, 26–35, Local visitor

> I may look upon objects in the future with a sceptical eye.
>
> Male, 26–35, International visitor

> I will pause for thought more often.
>
> Female, 26–35, Local visitor

Another idea that may be indicative of attitudinal change is realisation that there may be more to an object's physical state, or its history, than what is observable at first sight. It may be said that when visitors become exposed to the idea of potential interventions, they are likely to look for signs of them on museum objects and try to find out if and how an object has been treated.

> Because you can sometimes notice what has been restored.
>
> Female, 16–25, UK visitor

The Conservation Workshop at Yorkshire Museum was significantly divergent from the other studies in terms of the quantitatively measured impact on views of the condition in which museum exhibits are presented. Here all participants agreed there had been change in their way of thinking (100 per cent). The majority was confident that their participation experience had had an impact on their views, while the rest said that the workshop may have shifted their way of thinking. No one indicated that the workshop had no impact. By comparing these findings with the corresponding average data, it can be concluded that this workshop had the most notable measured impact on this strand of views. This extreme result could be explained by the double self-selection in the sample, as well as participants' social interaction with the author, both as workshop leader and researcher. The meaning of this striking consensus in the quantitative findings can be more fully interpreted with analysis of participants' qualitative justifications.

Similarly to the Conservation Galleries exhibition, the theme of suspicion towards interventions emerges from the realisation that objects may have been treated before they find their place in the museum's showcases.

> When I am next at a museum, I will consider how the exhibits are presented and whether I agree with what they have done. Before I wouldn't consider this at all.
>
> Female, 26–35, Local visitor

The workshop deliberately did not introduce any particular principles with regard to what constitutes good practice in conservation. This was decided to stimulate controversy and encourage argument among participants by introducing them to different historical and current trends in relation to conservation principles and ethics. Through their experience and the conclusions drawn from group discussion, most participants came to agreement that often there are no simple solutions in conservation, as there are no black-and-white certainties in approaching the construction of heritage. The realisation that conservation plays an important role in interpreting, and even shaping heritage, along with the meanings and values attached to it has resulted in some degree of scepticism and the development of a more critical outlook.

> I will think more about how it's done – whether I think it has been done as I think it should have been.
>
> Female, 51–64, Local visitor

As a result of the appreciation of the controversy between existing conservation theories, visitors developed an appreciation for preventive conservation approaches. Aspects of environmental control, as well as museological issues like labelling and interpretation, emerged through group discussion as an alternative to remedial conservation.

> [The workshop made me think about] lighting, storage, presentation etc.
>
> Female, 51–64, Local visitor

The experience has caused some to consider their views around the notion of perfection and physical integrity, which are not necessarily seen as desirable values in themselves. This is arguably a change of attitude rather than of a singular strand of views. For this visitor, meaning can instead be found in the appreciation of the different states of preservation, which is now understood as an integral part of an object's history, or, indeed, an object's biography.

> Do museum objects need to be exhibited whole or in perfect? I used to think so but perhaps its broken condition tells us more about its history.
>
> Female, 16–25, UK visitor

The conservation work at the National Trust's Knole House differed from the other case studies in the sense that here they were focused on providing visitors with an insight into different ongoing conservation projects in the collection, rather than introducing conservation as an exhibit in itself. The majority of participants responded that there was some change in their way of thinking, while more than half were confident that the exhibition made them think differently about the condition of museum exhibits, and some were ambivalent. Fewer said that the exhibition had no impact at all. By comparing these findings to average data, the measured impact of the engagement events at Knole House was substantial but slightly below average.

The qualitative responses demonstrate the kind of changes that occurred in visitors' views on the subject. The most prominent outcome recorded was, once again, the feeling of appreciation for the information provided, and the privilege of being given access to the processes taking place behind the scenes.

> Develop an appreciation of what goes on behind the scenes to preserve these artifacts for future generations.
>
> Female, 51–64, Local visitor

Conservation work is perceived as a painstaking and time-consuming undertaking that aims to ensure the protection of the collection. For most visitors, finding out more about it is seen as a way of understanding the conserved exhibits and their condition better.

> Helps understand the problems and appreciate the time scale and end product.
>
> Male, Over 64, UK visitor

It was also discerned that objects in the collection are subject to decay due to environmental conditions, which conservators try to keep under control. It was understood that certain aspects of the presentation of the collection, such as limited lighting and low temperature in the house, are connected to preventive conservation efforts.

> It made me understand how the environment alters the condition of objects.
>
> Male, Over 64, UK visitor

Finally, it was proposed that the presentation of exhibits should include information about conservation. This aspect was seen as part of present-day good practice.

> Exhibits should now be shown with conservation considered. I think most modern exhibits actually do this.
>
> Male, Over 64, UK visitor

How did the visitor experience impact on views about heritage conservation practice?

The quantitative analysis showed that the majority of participants (81.8 per cent) reported change impact on their views regarding heritage conservation practice, while almost half were very confident of this and some said there was no change at all (18.2 per cent).

In the Conservation Galleries, Ashmolean Museum, the majority of respondents replied that there was some change in their way of thinking, though most did not think that this was a major change, and only some were certain the exhibition

Table 9.3 Change in views on heritage conservation practice

	Ashmolean Museum (%)	Yorkshire Museum (%)	Knole House (%)	Average (%)
Yes, definitely	29.5	66.7	45.6	47.3
Maybe a little	56.4	14.3	32.9	34.5
No, not at all	14.1	19.0	21.5	18.2
Total	100.0	100.0	100.0	100.0

impacted on their views. Fewer said that the exhibition had no impact at all. Compared with the average results it can be said that the exhibition has had a greater overall impact on this strand of views than the average, but the changes that occurred were more subtly experienced by the visitors. The qualitative analysis showed that a significant proportion of responses were focused on the learning aspect of the exhibition (being presented with more information about museum conservation – a subject they previously knew very little about).

> Until visiting the galleries, I wasn't aware of the particular skills and thinking required in order to preserve particular objects from the past.
>
> Male, 26–35, UK visitor

> I wasn't aware how much work was involved.
>
> Male, 16–25, International visitor

Many responses were concentrated on the visitors' increased appreciation for the work of a conservator and its importance as a result of visiting the galleries.

> I could understand how conservationists work. Some things must take a lot of time.
>
> Female, Over 64, UK visitor

> Because [before my visit] I didn't appreciate the work done.
>
> Female, 51–64, UK visitor

> I hadn't thought about how important the right conservations are for preserving objects.
>
> Female, 26–35, Local visitor

Some respondents' discernment was widened as the exhibition revealed varied aspects of the objects and multiple dimensions to their history.

> It is very interesting seeing what is revealed about objects using conservation techniques.
>
> Female, 51–64, UK visitor

> The story behind objects is just as important as the object itself.
>
> Female, 26–35, UK visitor

> I also enjoyed the x-ray thing today, you know with the child, and the backlit paper. I've seen it before... it made me smile that things can be hidden either intentionally like the example of forgery, or that you can suddenly discover something new about something beautiful. So there is a beautiful old picture or painting and then suddenly something new is revealed about it. And that's a reason to smile I think. Like a hidden treasure, you know. Finding a

treasure... yeah! If you were the first person to look at it and you would have been looking at that painting for, I don't know how long, and then suddenly you put ultraviolet or x-ray, you would probably be running out of the conservation lab to call everyone 'Oh look what I found'! It's exciting I mean.

Female, 36–50, UK visitor

Those who indicated that there was no impact on the way they thought about conservation, mostly did so because they thought they already knew about conservation, or already appreciated it and understood its importance.

I don't think so, no. Because I already knew it was a very complex process, really scientific. So I don't think that it changed my views. It informed me better about how it works and the technicalities of it. It's really informative, really impressive.

Female, 16–25, UK visitor

I already understand the ageing effects that displaying objects can cause.

Female, 36–50, UK visitor

That's probably for other people, because I realise how important the conservation could be... So not really something new for my mind.

Male, 51–64, International visitor

The Conservation Workshop, Yorkshire Museum, was very impactful for the majority of participants but there were also many negative to change responses. The qualitative justifications for these responses contribute significantly to the interpretation of this data. A typical reason of a negative response is shown below.

I already believe that heritage conservation is extremely important.

Female, 51–64, Local visitor

Typical changes in the views of visitors who responded positively revolved around the themes of awareness, learning, deliberation over conservation ethics and principles and feelings of involvement in the heritage conservation process.

[The workshop] threw up lots of questions and concerns and made us aware of solutions or possible ways to look at museum exhibits.

Female, 51–64, Local visitor

I learnt a lot more about the questions and dilemmas involved and it made me think about whether access or preservation is more important.

Female, 16–25, UK visitor

I feel more involved.

Female, 51–64, Local visitor

Again, in the conservation work at Knole House the majority of participants replied that there was change in their way of thinking about conservation, but several also observed that the exhibition had made no impact on this.

> It is so easy to take things at face value and not think about how it is kept in such good condition. It is easy to take such things for granted.
>
> Female, Over 64, UK visitor

The vast majority of negative responses to this question can again be interpreted through the visitors' pre-existing notions that conservation is of great importance and/or of interest to many of them.

> I have always felt this work is important. It was interesting to see how much time (and money) needs to be spent on this work.
>
> Female, Over 64, UK visitor

> [Finding out about conservation work did not make me think differently at all about heritage conservation because] I was already interested.
>
> Male, Over 64, UK visitor

Conclusions

The analyses suggest that the events have had an impact on visitors' views on several different levels. Participants expressed a wide range of changes in their views regarding conservation as a process and the effect on their understanding of museum objects. Most agreed that the events studied have played a role in shifting their way of thinking. Participants had generally positive attitudes towards attempts of museums to educate them around conservation issues, and were overall very appreciative of the opportunity to find out more about an aspect of the museum that they, in their own words, previously ignored or had very little access to, both physically and intellectually.

Participants also showed signs of developing a critical understanding of the role of conservation in shaping the form and meaning of heritage objects while mostly expressing negative views towards restoration and more positive towards preventive conservation.

The findings of these analyses highlight the need for conservators and curators to consider visitors' views about many aspects of cultural heritage conservation, both as a social process and as a part of professional practice, if we are to devise effective strategies for communicating and negotiating its purposes together with society. As Lowenthal argues, 'history explores and explains pasts grown even more opaque over time; heritage clarifies pasts so as to infuse them with present purposes' (1985: xi). Conservation is thus another tool to shape values and notions already embedded in heritage, as Lowenthal defines it. Finally, as Swarbrooke observes, 'the reality of a product or experience is probably less important than the consumer's perception of it' (1996: 69).

In terms of quantifying the impact on views on heritage conservation practice, the findings show slightly less change. It was reported that participants already had an opinion about the importance of what conservation is and thus the events only reinforced that view, confirming Smith's argument.

The themes identified through the qualitative analysis were related to the educational value of having access to conservation, a sense of privilege – for being allowed to witness work behind the scenes of the museum, enrichment of visiting experience. At the same time, however, there was scepticism: some visitors said that they became more suspicious of the role of conservation as a result of their experience, causing them to either re-evaluate their views on the conservator's role or, in other cases, to reinforce pre-existing views/values.

The study presented certain methodological difficulties in applying the same methods across case studies, due to their different natures. One of the greatest challenges was the conservation professionals' tendency to seek affirmative measures of successful reception in positive feedback. One of the important lessons learnt in the process of developing evaluation strategies was that seeking evidence that an event has been 'successful' is not valid as a research objective unless an engagement project is evaluated against defined, intended outcomes. In addition, numbers of participants alone do not demonstrate impact. Collaboration with researchers to develop and apply robust, unbiased methods could significantly enhance an event's value and impact. In evaluation, it is more meaningful to use a combination of quantitative and qualitative methods to capture the multiple dimensions of the engagement process. Consistency is key in the choice of methods in order to achieve accessible and comparable findings that can inform future practice. Research-based guidelines for public engagement (such as the HLF evaluation guidance, Heritage Lottery Fund 2012) need to be followed. Visitor research should feed into institutions' and conservation practice to facilitate reciprocal reflection on experience rather than simply using one-way dissemination of conservation principles and/or techniques that commonly characterises public outreach. Finally, in order to achieve high impact and advance public engagement to the point of realising its promising potential for social benefit and economic sustainability, conservators need to consider the utility of reflecting on and learning from mistakes and failures, instead of persistently reporting success stories. We need to shift our thinking from the evaluation *of* success, and instead focus on evaluation *for* success. As suggested, referring to the usefulness of mistakes as empirical tools, 'sometimes you don't just want to risk making mistakes; you actually want to make them – if only to give you something clear and detailed to fix. Making mistakes is the key to making progress' (Dennett 2013: 20).

Note

1 Aas *et al.* 2005, Chirikure and Pwiti 2008, Damm 2005, DCMS 2015, English Heritage 2000, Hampton 2005, Hodges and Watson 2000, Kuper 2003, Marshall 2002, McManamon 2000, Smith and Waterton 2009, Strange and Whitney 2003, Townshend

and Pendlebury 1999, Tully 2007, Waterton and Watson (eds) 2011 and Watkins 2000 demonstrate the scope and growth in the last 15 years.

References

Aas, C., Ladkin, A. and Fletcher, J. 2005. 'Stakeholder collaboration and heritage management'. *Annals of Tourism Research* 32(1), 28–48

Arnstein, S.R. 1969. 'A ladder of citizen participation'. *Journal of the American Institute of Planners* 35(4), 216–24

Avrami, E., Mason, R. and De La Torre, M. 2000. *Values and Heritage Conservation. Research Report*. Los Angeles: Getty Conservation Institute

Bassey, M. 1981. 'Pedagogic research: on the relative merits of search for generalisation and study of single events'. *Oxford Review of Education* 7, 73–94

Bassey, M. 1999. *Case Study Research in Educational Settings*. Buckingham: Open University Press

Bennett, J., Rollnick, M., Green, G. and White, M. 2001. 'The development and use of an instrument to assess students' attitude to the study of chemistry'. *International Journal of Science Education* 23(8), 833–45

Berducou, M. 1999. 'Why involve the public in heritage conservation-restoration?'. In D. Grattan (ed.), *Conservation at the End of the 20th Century*. Paris: ICOM-CC, 15–18

Brooks, M.M. 2008. 'Talking to ourselves: why do conservators find it so hard to convince others of the significance of conservation?'. In *Preprints of the 15th Triennial Conference of the ICOM Committee for Conservation*. New Delhi: Allied Publishers, II, 1135–40

Brooks, M.M. 2013. '"Culture and anarchy": considering conservation'. In E. Williams (ed.), *The Public Face of Conservation*. London: Archetype, 1–7

Charmaz, K. and McMullen, L.M. 2011. *Five Ways of Doing Qualitative Analysis: Phenomenological Psychology, Grounded Theory, Discourse Analysis, Narrative Research, and Intuitive Inquiry*. New York: Guilford Press

Chirikure, S. and Pwiti, G. 2008. 'Community involvement in archaeology and cultural heritage management'. *Current Anthropology* 49(3), 467–85

Cooper, B., Glaesser, J., Gomm, R. and Hammersley, M. 2012. *Challenging the Qualitative–Quantitative Divide: Explorations in Case-Focused Causal Analysis*. London: Continuum

Council of Europe 2005. *Framework Convention on the Value of Cultural Heritage for Society*. Faro. Strasbourg: Council of Europe

Crooke, E. 2007. *Museums and Community: Ideas, Issues and Challenges*. London: Routledge

Damm, C. 2005. 'Archaeology, ethnohistory, and oral traditions: approaches to the indigenous past'. *Norwegian Archaeological Review* 38(2), 73–87

Dennett, D.C. 2013. *Intuition Pumps and Other Tools for Thinking*. New York: W.W. Norton & Co.

DCMS 2015. *Taking Part* [online]. Available at https://www.gov.uk/government/collections/taking-part [accessed 26 August 2015]

Eastop, D. 2006. 'Conservation as material culture'. In C. Tilley, W. Keane, S. Kuechler, M. Rowlands and P. Spyer (eds), *Handbook of Material Culture*. London: Sage, 516–33

Eastop, D. 2011. 'Conservation practice as enacted ethics'. In J. Marstine (ed.), *Routledge Companion to Museum Ethics: Redefining Ethics for the Twenty-First Century Museum.* London: Routledge, 426–44

Eastop, D. and Brooks, M.M. 2011. *Changing Views of Textile Conservation.* Los Angeles: Getty Conservation Institute

English Heritage 2000. *Power of Place: The Future of the Historic Environment.* London: English Heritage

English Heritage 2008. *Conservation Principles, Policies and Guidance for the Sustainable Management of the Historic Environment.* London: English Heritage

Glaser, B. 1992. *Emergence v Forcing Basics of Grounded Theory Analysis.* Mill Valley, CA: Sociology Press

Glaser, B. and Strauss, A. 1967. *The Discovery of Grounded Theory.* Chicago: Aldine

Hampton, M.P. 2005. 'Heritage, local communities and economic development'. *Annals of Tourism Research* 32(3), 735–59

Hein, G.E. 1995. 'The constructivist museum'. *Journal for Education in Museums* 16, 21–3

Heritage Lottery Fund 2012. *Evaluation: Good-practice Guidance* [online]. Available at http://www.hlf.org.uk/HowToApply/goodpractice/Pages/Evaluation_guidance.aspx#.U7OrlRb7UlI [accessed 5 May 2014]

Hodges, A. and Watson S. 2000. 'Community-based heritage management: a case study and agenda for research'. *International Journal of Heritage Studies* 6, 3

Hooper-Greenhill, E. 2008. *The educational role of the museum.* London: Routledge

ICOMOS 1994. *Nara Document on Authenticity in Relation to the World Heritage Convention.* Adopted in Nara Conference [online]. Available at http://www.icomos.org/charters/nara-e.pdf [accessed 17 November 2015]

ICOMOS 2000. *The Burra Charter: The Australia ICOMOS Charter for Places of Cultural Significance 1999.* Burwood, VIC: Australia ICOMOS

ICOMOS 2014. 'Heritage and landscape as human values: speakers and speeches'. 18th ICOMOS General Assembly and Scientific Symposium, Florence, 9–14 November

Jones, S. and Holden, J. 2008. *It's a Material World. Caring for the Public Realm.* London: Demos. Available at http://www.demos.co.uk/files/Material%20World%20-%20web.pdf [accessed 3 April 2011]

Kuper, A. 2003. 'The return of the native'. *Current Anthropology* 44, 389–402

Lowenthal, D. 1985. *The Past is a Foreign Country.* Cambridge: Cambridge University Press

Macdonald, S. 2002. *Behind the scenes at the science museum.* Oxford: Berg

Macdonald, S. 2005. Accessing audiences: Visiting visitor books. *Museum and Society,* 3(3), 119–36

Marshall, Y. 2002. 'What is community archaeology?'. *World Archaeology* 34(2), 211–19

McManamon, F.P. 2000. 'Archaeological messages and messengers'. *Public Archaeology* 1, 5–20

Merriman, N. 1991. *Beyond the Glass Case: The Past, the Heritage, and the Public in Britain.* Leicester: Leicester University Press

Muñoz Viñas, S. 2005. *Contemporary Theory of Conservation.* Oxford: Elsevier

National Co-ordinating Centre for Public Engagement 2012. *What is Public Engagement?* Available at https://www.publicengagement.ac.uk/ [accessed 15 November 2015]

Oppenheim, A.N. 1992. *Questionnaire Design, Interviewing and Attitude Measurement.* London: Pinter

Peters, R. 2002. 'Conservation as a later addition'. Papers from the Institute of Archaeology, University College of London, 13, 64–72

Pye, E. 2009. 'Archaeological conservation: scientific practice or social process?'. In A. Bracker and A. Richmond (eds). *Conservation Principles, Dilemmas, and Uncomforatble Truths*. Amsterdam: Butterworth-Heinemann, 129–38

Smith L. 2014. 'Theorising museum and heritage visiting'. In K. Message and A. Witcomb (eds), *Museum Theory: An Expanded Field*. Oxford: Blackwell Wiley

Smith, L. and Waterton, E. 2009. *Heritage, Communities and Archaeology*. London: Gerald Duckworth and Co.

Strange, I. and Whitney, D. 2003. 'The changing roles and purposes of heritage conservation in the UK'. *Planning Practice and Research* 18(2–3), 219–29

Strauss, A. 1987. *Qualitative Analysis*. New York: Cambridge University Press

Strauss, A. and Corbin, J. 1990. *Basic of Grounded Theory Methods*. Beverly Hills, CA: Sage

Swarbrooke, J. 1996. 'Understanding the tourist: some thoughts on consumer behaviour research in tourism'. *Insights* 8(1), 67–76

Townshend, T. and Pendlebury, J. 1999. 'public participation in the conservation of historic areas: case-studies from North-east England'. *Journal of Urban Design* 4(3), 313–31

Tully, G. 2007. 'Community archaeology: general methods and standards of practice'. *Public Archaeology* 6(3), 155–87

Waterton, E. and Watson, S. (eds) 2011. *Heritage and Community Engagement: Collaboration or Contestation?* London: Routledge

Watkins, J. 2000. *Indigenous Archaeology, American Indian Values, and Scientific Practice*. Walnut Creek, CA: AltaMira

Part 2

Case studies

Engaging conservation in community
practice

10 Living with history in York

Increasing participation from where you are

Lianne Brigham, Richard Brigham, Peter Brown and Helen Graham

Two of us – Lianne and Richard – are administrators on a Facebook page titled 'York Past and Present'. The title 'Past and Present' sums up what we've been working on together for the past 18 months. We've been trying to understand the ways in which York's heritage, as it is understood through photos, memories, buildings and collections, affects the lives of people living in York today and, more specifically, how more people can become actively involved in shaping decision-making about heritage and, through this, shaping decision-making about the future of the city more generally.

The four of us – with Paul Furness and Martin Bashforth – worked as a team on 'York: Living with History' as part of a UK-wide research project 'How should heritage decisions be made?' funded by the Arts and Humanities Research Council's Connected Communities project. Both the Heritage Decisions project as a whole, and the York: Living with History strand we are going to focus on here, have been the product of many minds and of many conversations. In the case of York: Living with History, these included conversations which took place at tens of public stalls, drop-in sessions, over many cups of tea and coffee, as well as at larger-scale events.

We thought long and hard about how to co-write a chapter for this book. The dangers of co-writing is that the person who has 'writing academic articles' in their job description (i.e. Helen) ends up taking on the job and the balance of voices and perspectives gets lost. Instead, we decided to honour our conversational method by recording our discussions and interspersing this with sections which to draw out key ideas. In doing this we also hope we can go beyond telling you how we've worked together but instead *show* you – model it – through the way the chapter is put together. Much as with increasing participation in heritage and conservation, increasing participation in research requires different forms and new ways of holding different ways of knowing together.

Conversation 1: living in York

HELEN Shall we start by talking about how we've come to live in York and care about York's history and heritage?

How should heritage decisions be made?

Over the last two years a team of 20 people – researchers, policy makers, funders, museum practitioners, people who are activists about their own history and heritage – have worked together to design and then carry out a research project.

The Heritage Decisions team were brought together by a pilot scheme developed by the Arts and Humanities Research Council's Connected Communities programme. The Connected Communities 'Co-design and Co-creation Development Awards' scheme sought not only to enable collaborative research between researchers, policy makers, practitioners and community groups, but also to actively enable the collaborative development of a research agenda, from its earliest stages.

While we all had a shared interest in heritage and decision-making, the team was formed deliberately to draw into dialogue people from different backgrounds, positions and approaches. The aim was to use the team's collective experiences, perspectives and positions to create a research project which might explore how to increase participation in heritage decision-making. One of the strands of work was the York: Living with History project.[1]

Who was involved in York: Living with History?

A core team of us led the York: Living with History project:

Lianne Brigham and Richard Brigham are founders of York Past and Present, which is a Facebook group – now with its own website – that has gone from three to 8,000 members in the last 18 months. Acting as a forum for exploring and sharing photos, the group shows the power of local knowledge and local networking and has crossed over into the real world through regular pub meetings and the 'public documentation' photography projects (which we will discuss further) at York's Guildhall and the Mansion House.

Peter Brown was Director, York Civic Trust. Peter first joined York Civic Trust to set up the Trust's Fairfax House in early 1980s. In 2011 Peter was awarded an MBE for 'helping to preserve the heritage of York'. He is author of *Views of York* (2012) and retired at the end of July 2015 to complete a number of book projects, including work on Georgian York.

Paul Furness is a radical historian and writer and led walks as part of the York: Living with History project titled 'York: A Walk on the Wild Side' (Furness 2014). As part of the project we made Paul's walks into a book, which now graces the shelves of many bookshops in the city (Figure 10.1). The launch of the book provoked debate about the implications of certain forms of heritage for York as a city in danger of losing its social and economic diversity as house prices spiral out of reach for anyone doing the minimum wage service jobs associated with the city's tourism industry (Lewis 2015a, 2015b; Graham 2014a).

Figure 10.1 Paul Furness published a book of his *York: A Walk on the Wild Side* walks as part of the York: Living with History project

Helen Graham is Research Fellow in Heritage, University of Leeds. Helen – along with radical family historian Martin Bashforth – has been involved in York's Alternative History since 2012. Before working at the University of Leeds, Helen worked in adult learning and community engagement in museums and heritage contexts – in the Heritage Decisions project she was able to bring all these different professional and activist connections together.

PETER The opportunity came up for me to apply for a job as the first Director of Fairfax House, which York Civic Trust had been restoring with a view of opening it to the public. I was appointed, which was great. I said I was going to stay for two years and give it my best shot; but 31 years later I'm still here in York.

RICHARD I've always lived here. Nothing really more to say. I just like York. Everywhere I've moved to I've always come back here. It's too nice a city to leave.

LIANNE Same with me really, although I've moved around everywhere in York, York has always been my home. I could never imagine being anywhere else.

HELEN I came here in 1995 to be a student at the University of York. So, I was an undergraduate in York. But I think the city almost acted like a backdrop to the intensity of student life. So, I think it was only when I started to study for my PhD – which was on the Women's Liberation Movement in 1970s and 1980s – and, at the same time, live in the city centre that I started think seriously about the city.

Conversation 2: getting involved in Heritage Decisions project

HELEN Peter, I remember writing you an email about getting involved in the Heritage Decision project.

PETER Yes. Working for what can only be described as a traditional organisation, populated by old fogies, it seemed a really good idea to get some fresh ideas and some fresh blood circulating through my veins. And your offer to be a collaborator seemed very exciting. The project became more effective and more interesting as time went by; especially in those first engagement sessions where we were really trying to drill down to: what exactly what do we mean by heritage? What do we mean by value and all these issues? And what right does that expert have to make decisions on our behalf?

HELEN Out of those early conversations – part of the co-design process – we drew out a research agenda which pushed some of these questions about how to involve more people within decision-making. And out of this came the York strand of the project. We had quite ambitious research questions: about trying to understand more 'how heritage affects the lives of people who live in York', but with the aim of using the energy around that question as a way of thinking about how we could lever open space for greater participation.

In terms of first meeting you, Lianne and Richard, I'd been on Memories of York Facebook page for a while, which was very popular. I think even in 2012 it had 11,000 likes; there were lots of people on there. But it was also quite a tense space. There was quite a lot of quite angry disagreements going on through Memories of York. It was only just at the start of the York: Living with History project I became aware of a new Facebook group: York Past and Present.

LIANNE We first met 27 February 2014 at the Bike Shed Café.

RICHARD What I liked about it was you were doing something very similar to what we wanted. The point is we've got annoyed so many times with wanting to do something in York with buildings rather than change, demolish, pull down whatever. And you can't seem to break through that barrier to get through there. You say, 'I want to do this' and you're just looked at as a member of public. It's like it doesn't really matter what you want. That's something we wanted changed. We wanted to see whether we could actually get enough people involved to make an impact. It's different saying me and my wife want to do it; but when you say there are 8,000 people you are working with…

Act: make change from where you are

The Heritage Decisions project identified four key ideas, which we will introduce throughout this chapter. The first was 'Act'. Our initial research in York suggested that one of the greatest barriers to participation was the strong feeling – often based on concrete experiences of disempowerment and exclusion – that it isn't possible to get involved and influence things. One thing any of us can do to refuse this sense of powerlessness is to try and make things happen. York Civic Trust, York Past and Present and York's Alternative History are, in their different ways, all examples of this. We liked the idea of 'making change from where you are' – not waiting for someone in a more powerful position to initiate, but using the agency you do have to make something happen.

Conversation 3: is it possible to make a difference?

LIANNE When we first had the meeting at the café, although it was something that we wanted, I seriously thought that we were all just getting high hopes that we'd actually be able to get somewhere; that we'd actually be able to chip away at the Council, chip away at the authorities. So, I did come away, not disillusioned, but a bit cynical to be honest. I did think, this isn't going to happen, this isn't going to work. But looking back, good things have happened. Because for us as a group it's opened up so many opportunities that I don't think we would have been able to get. And working as a team we've opened up a lot of people's eyes that actually we're not going to go away; we have a voice and we can be heard now.

PETER You've suddenly become legitimate.

LIANNE Yes, exactly. Not just two people.

Figure 10.2 As part of the Heritage Decisions project, Danny Callaghan, who works on the
heritage of the potteries in Stoke-on-Trent, developed a 'DIY Heritage Manifesto'

RICHARD It's allowed us to meet people like John Oxley (City Archaeologist),
Richard Pollitt (Mansion House, Guildhall and Civic Services Manager),
yourselves (Peter and Helen); people we wouldn't normally meet. So, if we
have a question we have these different people that we can contact. I think
that's probably the most important thing is the different people you have to
meet that will help you in every different direction.

PETER But it's a two-way process. There is no way that we would, as an organisation, have the resources to document the Mansion House, for example. But you have committed a huge amount of resource to documenting to the minutest detail. But you're documenting the Mansion House and its changes and you're recording them for posterity. That's fantastic.

HELEN Lianne, can I just ask a bit more about the cynicism that you felt? Because it wasn't an unusual reaction to speaking to me! Most of conversations I had in first few months doing the project – all with people who weren't professionals – were with people feeling that decisions were generally made behind closed doors and that decisions are often made long before they come into the public domain. There was a real sense of cynicism about the political culture of the city really. I don't think that's necessarily just about York; I think it's to do with the political culture of the United Kingdom more generally. But obviously in York it was expressed in relationship to the Council. So, I don't think what you were thinking was unusual at all… it was a common theme in most of those conversations.

LIANNE We had had a specific experience that made me feel like that. We had tried to photograph the Hutments behind York Art Gallery before they were demolished. Literally for three weeks we'd been passed from pillar to post. All we wanted to do was document them before they went.

PETER Without putting words in their mouth, they were probably frightened that you were going to mount a campaign to have the Hutments saved (see Figure 10.5).

LIANNE (Laughter) Possibly.

PETER Maybe a conversation might perhaps have put their fears at rest.

RICHARD We did try to talk to them. At one point I did talk to a person at the museum, I can't remember who it was, and I did say, 'We understand that it's going to be pulled down. We just want it documented.' But it was early days and when you don't know anybody, it's very difficult. We find now that having your foot in the door makes everything a lot easier than being completely out of it.

LIANNE The issue with the Hutments behind the art gallery was before we started York Past and Present – it was one of the reasons *why* we started York Past and Present. So, that was an experience for us – maybe not a good one. Then when we started talking to you, Helen, you were wanting to work and do things similar to what we were wanting to do. There was still a lot of cynicism because we weren't too sure we would get anywhere.

RICHARD It's been like a big machine: we knew where we wanted to go, we just didn't know how to get there, and you've been like our SatNav; you've showed us which direction to go in. You have. You've pointed us in the direction of how we can do it. And all the other people we've met, just different branches, it's worked really well.

HELEN But SatNav makes it sound like I knew what the direction was or had a sense of the territory. But I didn't. The map of the territory has been something we've been building together. And we needed all the different viewpoints on it otherwise we wouldn't have been able to see what we've seen. It's like we all had little vantage points on the territory; but actually it's by bringing them together that we've developed a richer understanding of how decision-making works within the city. And then also some of the potentials for opening it up too.

Connect: cross boundaries and collaborate

At the heart of the Heritage Decisions project from the first was the idea of diverse perspectives across hierarchies and institutional boundaries. The York: Living with History project worked to build a local network in the same way. The aim of this was both to develop a much richer understanding of 'York' and 'heritage' than could be gleaned from any one perspective. But also, through working with people with different scopes for action, we could also identify ways of creating new ways of increasing participation (in this we were influenced by Gilchrist 2009).

Richard Brigham, in the project's booklet, *How Should Heritage Decision Be Made? Increasing Participation From Where You Are*

> We've found that networking works. There's like this magic path. You need to find one person and then they introduce you to their friends. There are two types of people in the council/organisations. The ones that want to work with people and want change; and those that don't. The key is find those that do want change and then they usually know other people who do too.
>
> 2015: 6

Conversation 4: mapping heritage decision-making – and then seeking to intervene

HELEN One of the things that was a bit eye opener for me was... I think from the outside you can think that the Civic Trust – and it does have influence –
PETER It has some.
HELEN Yes. And certainly the Civic Trust is part of the official consultees for various decisions that will affect the heritage of the city. At the same time it's not like you necessarily have as much influence as people might think.
PETER We don't have the final say.
HELEN Exactly.
PETER I'd love to. (Laughter)
HELEN Seeing actually how even influential organisations also face some of the same issues that individual members of the public do was part of the work that we were trying to do.

Situate: understand your work in context

The Heritage Decisions project used 'thinking systemically' as a research methodology. Mapping heritage decision-making systemically proved powerful. If you can see how formal structures and informal networks fit together, then you can start to notice key people and key points for increasing participation in decision-making. We also found this technique useful for reflecting on our own practice and activism and for planning action and making connections.

Thinking systemically

Systemic thinking offers a way of thinking about heritage not in isolation or fixed but as a dynamic process which is produced, and shaped, by people, ideas and things, and the way they interact – and don't.

- Map processes
- Look for patterns
- Notice boundaries and disconnections

We were particularly inspired by Systemic Action Research pioneered by Professor Danny Burns in a Development Studies context – an approach which refuses to see 'knowing about' and 'change' as separate or sequential processes. Danny Burns (2007) shows the importance of drawing on lots of different people's knowledge within a local system to develop a 'working picture' and to recognise that 'each situation is unique and its transformative potential lies in the relationships between interconnected people and organizations' (ibid.: 32). The key point there is that neither 'power' nor 'change' is never abstract, it is a capacity of the people within any given system, organisation or locality – so if you're thinking about a city then it's crucial to try to develop a 'working picture' with the widest and most diverse set of people possible and actively involve key people in generating new insights – because it is through this process that common understandings and potential for change are cultivated (Edwards 2012).

Case studies: Stonebow House and Castle and Eye of Yorkshire

As part of the York: Living with History project we did more sustained case study work around two areas about which future decisions might be made. One was Stonebow House, a 1960s brutalist building which provokes a wide range of reactions in the city – most commonly hatred and ire (Graham 2014b)! One of our aims with the Stonebow project was to explore how to enrich the public debate

Figure 10.3 Stonebow House: an aim of the York: Living with History project was to see if we might enrich public debate through putting different people in dialogue with each other © York Mix

about Stonebow House – not necessarily to change anyone's minds but to create a context where better arguments, both for and against the building, might be developed. As part of this, Jon Wright, freelance architectural historian and former Senior Caseworker for the Twentieth Century Society and, until 2013, Head of Conservation at the Council for British Archaeology in York, shared his position on Stonebow House, in writing and at an event, and also engaged in sustained debates through York Past and Present (Wright 2014; Graham 2014c, 2014d). Through this process the quality of the public debate did develop. People who hated the building developed more sophisticated arguments to support their position. Some people changed their minds, one participant saying 'it's a bit like the more you interact with it, the more friendly it becomes'. We used the process to map the arguments, concluding that while there was no killer argument for keeping it, there were lots of small arguments (that seemed to add up for people) for exploring reuse and keeping it as an affordable enclave for shopping and socialising in a city becoming increasingly expensive (Graham 2014d).

The second case study was Castle Area and Eye of Yorkshire, a strand led by Peter and York Civic Trust.

PETER In the past, to go back to 1990, when the council had the idea for comprehensive redevelopment of Piccadilly and the castle car park, they produced a development brief, which we agreed with. We thought the ideas of massing, intensity of development and the number of people living in that site seemed fairly reasonable actually. So, that was given out to a number of firms, and they all came back with proposals for the redevelopment of the site. But it was as if they threw the development brief in the bin – the designs came back with twice the intensity, to six storeys instead of four. All the schemes were thrown out.

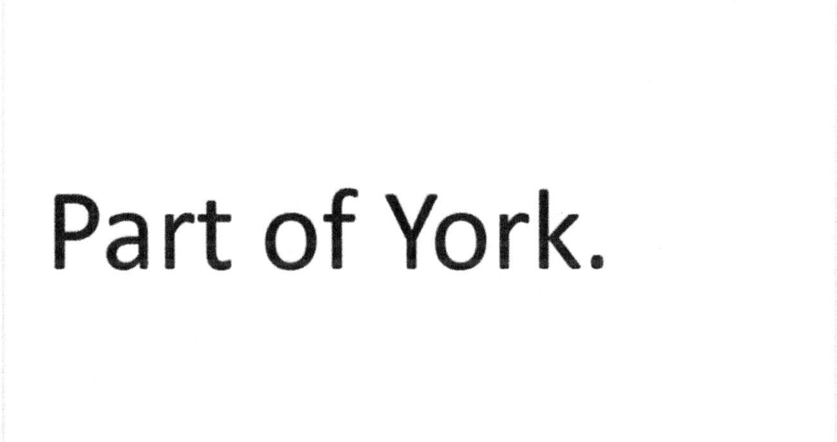

Part of York.

Figure 10.4 One of the Stonebow House in three word contributions generated by York Stories (2014) and at the research events

By the end of the 1990s the City Council were starting to lose the will to live, as were we, and they were desperate to try and get something happening there. York really felt it had to push on the development.

There was this sense that 'We have to get this huge department store done at the bottom end to draw people down'. As a result, the scheme that was agreed was the worst scheme that we had ever seen. The City Council approved it, officers had approved it; even the senior guy at English Heritage agreed with it. Anyway York Civic Trust put our money where our mouth is and we said, 'We really want this called in for Inquiry,' knowing full well that this was going to cost us a fortune. Seven months of my time, as it was, plus all my other trustees, £88,000 it cost us for the barrister, a seven-week Inquiry, and we won.

Since then the whole thing has been in abeyance. But we're now starting to see this idea creeping back in; and we just felt it was a very good opportunity to actually engage in this grassroots approach. What do we want to see on this site? What's right for the site? What's right for the city? So, we had people from 36 different organisations, some formal, some informal. We spent a day with a really good moderator by the name of Graham Bell from North of England Civic Trust, who was then able to draw that into a sort of coherent document which can be used and can be considered by people.

HELEN That history of activism is a crucial thing to remember: that it really was people, the Civic Trust, but also a whole range of individuals that stopped it.[2]

PETER Yes. There were about 40 or 50 in a group. We even had CABE, the Commission for Architecture and Built Environment, and they sent a barrister up as well. The developers had one of the top QCs in the country; the City Council had another top QC. They must have spent millions of pounds on it. Okay, we spent £88,000; but CABE must have spent hundreds of thousands plus. People gave so much time and effort.

HELEN There is a history in York of schemes coming forward supported by the Council, which were stopped by people in York, various activists and networks. The other example is the Inner Ring Road, which would have seen lots of houses knocked down.

PETER Good houses.

HELEN And – at a similar time – also lots of multi-storey car parks built as well.

PETER Four.

HELEN Four of them, yes. That was a campaign. Was that York 2000?

PETER York 2000, yes.

HELEN So, there is a history of people living in York refusing certain types of developments and making those cases, which I think is an important history for us to remember and gives a bit of context and tradition to some of the things we're trying to do today, I think.

PETER The shame is, of course, it's always adversarial.

HELEN Yes.

PETER So, how do we devise a system where we're not always in conflict and we're not always reactive? How do we be proactive and democratic right at the beginning? Absolutely crucial.

LIANNE I think the Castle Area and Eye of Yorkshire event was an eye opener for myself and Rich. We walked away thinking, well no wonder, because it was all about how much red tape you have to go through to get planning permission to even put in an application, let alone get it approved or anything.

RICHARD It would be nice to have more. If you could find a way to involve the public more directly in planning permissions and understanding the way they work then I reckon you'd get a next generation of people who actually understood how it all worked. I think that's important.

PETER That's a very good point.

Public partners: collaboration is key

The Castle Area event was written into a report by Graham Bell (2014). The key finding of the event, outlined in detail in the report, was that a project brief was needed, aimed at bringing together the three key partners (City of York Council, English Heritage and York Museums Trust) in a way that could fully address the physical improvement of the car park, the Eye of Yorkshire, the Foss and other public areas and ultimately coordinate management of this area. Such collaboration – with strategic and ongoing community involvement – was seen as the best foundation for doing justice to the complex and highly significant histories of the site.

Conversation 5: personal memories/public documentation/political decisions: linking different scales

LIANNE Looking back it seems so long ago. At one of the research Drop-In sessions at the Library we got talking to John Oxley, City Archaeologist. We were saying, basically, 'These buildings are getting changed or altered or demolished or whichever way you want to look at it, but they're not getting documented properly.' We wanted to do Urban Exploration – but with permission. So, it was then that we came up in the idea of: why not ask the public? There are lots of people with lots of cameras; why not ask them to go in and take photos and at least then you know that you're going to get a proper documentation.

HELEN You've taken 900 people around the Guildhall – a direct result of that conversation?

LIANNE Yes.

HELEN The aim has been that people have been able to get in. You talk a bit about each room that you're in, don't you?

RICHARD Yes.

HELEN But then people get to take their own photographs and then talk about the things they know about it as well?

RICHARD The weird thing is that the places that are of least interest, like the basements, are probably the parts that more people enjoy. You show people the

Figure 10.5 Taken by Richard Brigham as part of the pilot public documentation projects near York's Guildhall © Richard Brigham, York Past and Present

main hall and they go, 'Wow!' You take them into the basements and they're like kids with new toys; they're running around all over. Because they never get to see that. And that's what it is: everybody wants to see something they've never seen before or they're not allowed to see.

HELEN What's fantastic is you are building this understanding about the building, not just through books, but also through all the stories that people share on all of the tours.

RICHARD I think the stories, whether they be work stories or memories you have as a kid or anything that's personal to you, is significantly important to any building; much more than what's written down about its history, because now it becomes more personal. And I think a lot of people who work in buildings can tell you a lot better stories about say Guildhall, the Mansion House, the Fairfax House, than the people who write about it. You probably see some really weird and wonderful things in old buildings, but nobody actually gets round to write it.

HELEN Linked to that, it's interesting on Facebook how the personal and the city as a whole become intertwined.

RICH Yes.

HELEN So, someone will post a photo that is a family photo, but then lots of other people will post comments.

RICHARD Yes, it's happened loads of times. It happened a couple of months ago: somebody posted a picture of a bunch of people from Rowntree's in about '55, '60, I think it was. And then somebody said, 'It's my dad stood next door' and the other one was like, 'Yeah, he went to work at Terry's.' Then somebody else posted a picture of him at Terry's and he went, 'That's my old…' and it went from there. All these different establishments and all these different people, it became more like a history of chocolate and people who worked for them than anything else. It was great.

LIANNE We also found out, on our YouTube channel we've got a video of the railway, and one of our members came up to us the other day and said, 'I really want to say thank you for that video.' I said, 'Why, it's just a video?' He said, 'Because before I'd seen that video I'd never seen my grandad before.'

Personal contributions and political participation

If one of the things we need to change to increase participation is the widespread sense of powerlessness, then finding the pathways from personal and local knowledge and expertise and political participation is crucial. We've noticed (and tried this out explicitly with Stonebow House) how involving people in sharing memories and knowledge about a particular building, street or issue in the past – and being exposed to others' knowledge – both enables for us all a rich and more complex appreciation of the issues, but also offers a way in for more active engagement with decisions today.

Conversation 6: legitimacy?

HELEN One of the things we've been doing in York: Living with History – and on York Past and Present you've been doing it anyway – is trying to build broader networks, more diverse networks throughout the city of people. The event we held on 20 June 2015 'What has heritage ever done for us?' – which was part of the Arts and Humanities Research Council's Connected Communities Festival event – gave us a chance to bring everyone we've met together.

RICHARD Yes.

PETER I thought it was an excellent affair. You talked about all the work that's gone before that helped to legitimise the questions the project has been asking – and its methods too. I think you had a really interesting and diverse range of people, who wouldn't necessarily have engaged in that process in the past. So, that's a fantastic first step. We just need to work out how we can keep that momentum going: how do we prevent it just fading away? How we've got to convince the decision makers in the city that there is real benefit in community engagement; that's the challenge, isn't it?

HELEN You said earlier on about legitimacy. When you say that, what kinds of things do you mean?

PETER I mean, for example, I wouldn't have thought that [everyone who spoke] would necessarily have been involved in the 'What has heritage ever done for us' meeting if they hadn't been involved in other components and parts of the build-up to that meeting. So, that's really what I mean. I wasn't suggesting that we were illegitimate before. I'm suggesting that there is now more credence – more veracity.

HELEN And we've done that through conversations – and building relationships – with one person at a time. You're right, by building a broader group network then you do grow and become more recognisable and, in a sense, more legitimate. And you can tell you are growing in legitimacy by the fact that a wider range of people are prepared to speak to you or take part in events you're doing.

PETER Yes, when they realise that you're talking sensible words rather than… You can be radical and still be sensible.

LIANNE It's about people feeling empowered to know that they can speak up. We have a valid opinion.

RICHARD I think there must be loads of people out in York who think: my voice doesn't matter because what am I going to do, I'm just a person? We've all done it; we've all said we can't make a difference. But by joining everything together we are making differences. It might be small and small steps but it is a change. And if you can keep that going and progressing then…

LIANNE It's exactly what you were saying Peter about not letting it just fall. We've had so many highs, we need to keep going. We need to keep going so it doesn't become stagnant; it doesn't all just fall down and end up being for nothing.

PETER Return to the status quo.

LIANNE Yeah.

Reflect: seeing your work through other people's eyes

One of the most powerful outcomes of the wider Heritage Decisions project – and its collaborative design – was the chance for us to reflect on our own work and become more self-conscious about our approaches and choices. This was made possible through individual conversations between team members and the powerful effect of seeing our work afresh through other people's eyes. In York this was even more intense as we were constantly having to adjust our understandings as we saw issues from different perspectives. As part of the project Helen has kept a Social Network Analysis of all the people she spoke to. This was a way of capturing conversations and tracks how some relationships develop and gather pace and others falter; it was also a way of checking the diversity, or otherwise, of the networks the project was developing. The question has clearly become – how do we embed these methods of reflection and diversity in the political culture of the city?

Conversation 8: beyond an adversarial political culture

HELEN Peter you said earlier on about it always getting adversarial. I think that's a really crucial point.

PETER Yes.

HELEN The worst thing about that is not only that there might the loss of something that lots of people care about – such as a building – but more that actually, yet again, people feel like: I can't make a difference; there's no point. So, there are all these negatives in terms of the effect on political culture, and the cynicism will just grow if people's experience of getting involved in something like that is they don't get to be listened to and can't make a difference.

At the 'What has heritage ever done for us?' event there was a sense that we need to be proactive about getting people involved in what's important to them about the city – but also that we could maybe use history and heritage to enliven and create more fun and dynamic ways of people engaging in decision-making about the city more generally.

RICHARD Being proactive is partly about confidence. I think each group grows as well, the more networks you are connected into. Each network you're with seems to give you a little more confidence. It's like when we first talked to you (Helen) I didn't really want to say too much because it's just me at the end of the day. What do I know? You're all professionals. But as we've gone forward and met different people there's that build-up of confidence as well that you have got a voice and you will be listened to. And that's important.

LIANNE You can argue your case to us; we can argue our case to Peter; Peter then can argue his case to you – but you can argue your opinion with each other politely and still walk away at the end of the day and go, do you know what that was really fun, it works. It's taken in a more productive way that says, yeah, you feel more empowered; you feel more able to take out these points of views and put it forward and say this is what we want.

RICHARD Everybody thinks their point of view is right, till they hear somebody else's point of view that is actually better than theirs, and then it's an agreement.

PETER For the strategy for the future I think somehow all these various tentacles need to come together but still retain their independence. *How* do you do that?

As part of the project Lianne and Richard developed some advice to other community groups starting out:

- build a community of people with the same interest;
- meet key people based in institutions (like in a university or in the council). 'Be willing to give your free time because it's volunteering that makes the connections';
- tap into these key people's networks;
- cross over from social media to real life events and real life events back to social media;
- you need an idea to talk about [for York Past and Present this has been public documentation];
- communication and language – you need to speak to different people in different ways;
- gain more and more confidence from speaking to lots of different people – and seeing that they like what you are doing;
- be generous to people in institutions/council. 'They can shut the door, faster than you can open it';
- don't settle for small things, keep your eyes on your main goal. 'Be like a child, be happy with what those in decision-making positions offer but always demand more'.

Still living with history....

One thing we've learnt from the ways of working we've used in York: Living with History – and in the Heritage Decisions project more generally – is that, unlike traditional research projects, it won't ever be 'over' as such. It is often said in Heritage Studies that heritage is a process (e.g. Smith 2006). Equally, democracy is often described in the same way. Not a cross in a box 'every five years' (to paraphrase Tony Benn), but as a way of living together that needs to be constantly nurtured and nourished through working together across inequalities and boundaries (e.g. Dunn 2005). The Heritage Decisions project has tried to find a way of working which is adequate to both these conceptualisations. We've approached research and knowledge production in a way which is itself living and adaptive – always changing as it seeks to create the social conditions and capacities for participation in decision-making in York to be increased. All of the four key ideas from the overall Heritage Decisions project – Act, Connect, Situate and Reflect – always need to be maintained and are never completed actions. You always need to add and enrich your understandings of the complex contexts you are working in. You always need to diversify your networks. What we hope we have shown that it isn't helpful in the debates over heritage and communities to create and reinforce polarisations. It is not always helpful to think about professionals and communities. Or an inside and an outside. Instead, we can all grow networks of people across boundaries who can constantly challenge you to think differently and refuse to let you off the hook and think that nothing can be done.

Notes

1 For more information on the Heritage Decisions project see http://heritagedecisions.leeds.ac.uk/
2 Activists in the York Castle Area Campaign Group have donated their archives of the campaign to York City Archives in July 2015.

References

Bell, G. 2014. *The Castle Area and Eye of Yorkshire: A Gathering of Ideas.* Available at https://livingwithhistory.wordpress.com/2015/01/19/the-castle-area-and-the-eye-of-yorkshire-a-gathering-of-ideas-report-and-forthcoming-events/ [accessed 8 July 2015]

Brown, P. 2012. *Views of York.* York: York Civic Trust

Burns, D. 2007. *Systemic Action Research: A Strategy for Whole System Change.* Bristol: Policy Press

Dunn, J. 2005. *Setting the People Free: The Story of Democracy.* London: Atlantic

Edwards, A. 2012. 'The role of common knowledge in achieving collaboration across practices'. *Learning, Culture and Social Interaction* 1, 22–32

Furness, P. 2014. *York: A Walk on the Wild Side.* York: York's Alternative History. Available at http://yorkalternativehistory.wordpress.com/ [accessed 8 July 2015]

Gilchrist, A. 2009. *The Well-Connected Community: A Networking Approach to Community Development.* Bristol: Policy Press

Graham, H. 2014a. 'Is there a better way to make decisions on York's history? Join us to find out'. *York Mix*, 12 March. Available at http://www.yorkmix.com/life/people/is-there-a-better-way-to-make-decisions-on-yorks-history-join-us-to-find-out/ [accessed 8 July 2015]

Graham, H. 2014b. 'Stonebow House: locals have their say – "It's about the content, what's inside"'. *York Mix*, 10 April. Available at http://www.yorkmix.com/life/environment-life/stonebow-house-locals-have-their-say-its-about-the-content-whats-inside/ [accessed 8 July 2015]

Graham, H. 2014c. 'Stonebow House: Not exactly "YES" or "NO"... at our last event there was a lot of grey'. *York: Living with History*, 6 August. Available at https://livingwithhistory.wordpress.com/2014/08/06/stonebow-house-not-exactly-yes-or-noat-our-last-event-there-was-a-lot-of-grey/ [accessed 8 July 2015]

Graham, H. 2014d. 'Stonebow House: Running the Arguments'. *York: Living with History*, 6 August. Available at https://livingwithhistory.wordpress.com/2014/08/06/stonebow-house-running-the-arguments/ [accessed 8 July 2015]

Heritage Decisions 2015. *How Should Heritage Decision Be Made? Increasing Participation from Where You Are.* Available at http://heritagedecisions.leeds.ac.uk/ [accessed 8 July 2015]

Lewis, S. 2014a. 'Is York "horribly twee and as dull as ditchwater"?'. *York Press*, 26 January. Available at http://www.yorkpress.co.uk/news/11748834.Is_York___horribly_twee_and_as_dull_as_ditchwater____/ [accessed 8 July 2015]

Lewis, S. 2014b. 'So you thought you knew York? Think again...'. *York Press*, 26 January. Available at http://www.yorkpress.co.uk/features/features/11749189.Prostitutes__thieves_and_revolutionaries___York_s_alternative_history/?ref=rss [accessed 8 July 2015]

Smith, L. 2006. *The Uses of Heritage.* Routledge, London

Wright, J. 2014. 'Stonebow House reminds us that York lived through the 20th century'. *York Mix.* Available at http://www.yorkmix.com/life/environment-life/stonebow-house-reminds-us-that-york-lived-through-the-20th-century/ [accessed 8 July 2015]

York Past and Present 2013–16. http://yorkpastandpresent.wix.com/yorkpastandpresent [accessed 8 July 2015]

York Stories 2014. 'Stonebow survey results', 7 August. Available at http://yorkstories.co.uk/stonebow-survey-results/ [accessed 8 July 2015]

11 Community heritage and conservation in Castleford, West Yorkshire

Alison Drake

A perspective from the community, presenting some of the issues faced by Castleford Heritage Trust in working to reclaim, use and celebrate our local heritage in changing times and landscapes.

Introduction

After 15 years of working to deliver a wide variety of community heritage projects, as we all attempt to manoeuvre through the changing demands of these difficult times, Castleford Heritage Trust (CHT) has embarked on its most ambitious and demanding project to date. Castleford Heritage Trust purchased Queen's Mill on 17 April 2013 (Figure 11.1). It is the world's largest stoneground flour mill and was previously known as Allinson's, being the home of Dr Allinson's famous whole-wheat, stoneground flour, which has the 'Wi' nowt taken out' catchphrase.

In these days of austerity and shrinking budgets some may think it a foolhardy venture, but to our community it is a necessary act, to try to conserve, use and own at least one tangible aspect of our varied, unique industrial and social heritage. We will need to work in partnership to achieve our aims and vision for the development and use of Queen's Mill as a high quality community facility providing a unique heritage and arts destination for local communities and visitors from across the region.

We see this project as a major vehicle in achieving our longstanding aim to reclaim and use our tangible and intangible heritage and culture to build a better life for our community. Alongside a quality regional arts and heritage venue, we wish to develop access to the river as a recreational resource in the centre of town, to help develop and promote Castleford as a thriving market town. We hope to renew a community that has been struggling with high levels of social deprivation since the loss of its five coal pits and wider de-industrialisation in the second half of the twentieth century.

In this chapter I will give some background to the establishment of the CHT and then examine some of the issues and problems we have faced over the years

Figure 11.1 Queen's Mill as the former Allinson's Mill, purchased by Castleford Heritage Trust 2013

in delivering community heritage projects in Castleford, including the purchase and development of Queen's Mill so far.

Castleford Heritage Trust

Castleford Heritage Trust (CHT) was established in 2000 and is a registered charity and a limited company, which aims to promote the town's heritage and culture to build a strong, successful community. We use natural as well as cultural heritage as a vehicle for regeneration and improving educational experiences and opportunities.

CHT seeks to improve the local area and knowledge of local heritage in order to improve community esteem, raise the aspirations of our young people and for the general benefit of local people and visitors to Castleford. We have promoted access to our historical and archaeological collections, our waterways, ancient woodland, parks and other green spaces. We have developed educational opportunities and community involvement in natural and industrial heritage projects and celebrations.

The Castleford community came together to form the CHT at a time when the new millennium and the New Labour government's 'Agenda for Change' gave us hope and renewed drive to address some of the problems of our community. These problems had emerged during the 1984–5 miners' strike and the extensive and brutal pit closures programme of the Thatcher government. As a town of 40,000 people we were still unable to overcome the effects of losing a wide variety of mining and manufacturing industries throughout the second half of the twentieth century. At the turn of the millennium the former coalfields areas were struggling more than most.

At that time the government Social Inclusion Unit came into being to help government departments identify and address social inclusion problems. This requirement to act extended to the publicly-funded heritage sector, as it was successfully argued that heritage was well placed to contribute to building community and social capital, improving social inclusion and community life. Tensions developed between government aspirations and policies and some heritage sector professionals. However the response of the Department for Culture, Media and Sport (DCMS), the government department leading on heritage issues, was that 'Every library, museum, gallery and archive can recognise its responsibilities and should be considering and reaching informed decisions about how it can best meet the needs of its communities' (DCMS 2001).

The New Labour government's team for implementing policy in this field, Policy Action Team 10 (PAT 10), reported on the contribution arts and heritage can make in neighbourhood renewal. Heritage professionals were required to consider new theories in political philosophies of 'civil renewal', and management of 'community heritage' required a practical approach to improving quality of life, involving local people in decision making, and identifying and solving what communities viewed as problems for their areas.

Castleford people were ahead of the game. They were already mobilising to use local heritage to motivate community participation in consultations and actual delivery of regeneration projects using community heritage. The work has never stopped and, providing there is funding to keep us going, we will continue to deliver for our community.

Even now, all these years later, an enlightening study has been commissioned by the Coalfields Regeneration Trust, researched by Sheffield Hallam University, and recently presented to government. The report, *The State of the Coalfields* (Foden *et al.* 2014: 5–8), shows that:

> the disadvantage in the coalfields is deep-seated rather than rooted in the recent recession… the core of the problem goes back… even before the pit closures… The job losses of the 1980s and 90s still cast a long shadow… and evidence provides a compelling case that most of the coalfield communities… require support.

So our work is no less required now than when we first began and we know the development vision for the Queen's Mill project is vital to the future of our town and community.

Working with partners: issues and lessons learned

Much of our work over the years has involved partnership working, with a wide variety of organisations and groups, regional agencies, local government depart-ments and national bodies, such as the Homes and Communities Agency, to deliver projects and programmes. These have included major urban regeneration and environmental improvement initiatives, community arts and activity programmes, heritage skills and training workshops, access to nature and ancient woodland management initiatives. All this has been part of what we consider to be essentially 'heritage work', as it has involved using what we have inherited from the past to benefit our people today and to build a better future.

However, not all organisations and bodies are easy to work with and this is not always because of difficult individual personalities. The issues can stem from a variety of sources: sometimes it is easy to see the reasons for tension or difficul-ties and sometimes it is not.

A variety of issues and difficulties have emerged in the many community herit-age initiatives we have undertaken, but also in the day to day activities of CHT This chapter sets out some of these difficulties below, outlines the reasons for them and also recounts some of the lessons we learned. The difficulties stem in large part from the different agendas that professionals bring to the projects, as they attempt to work with the community in what are invariably unequal partnerships.

Some organisations are better than others at adapting to the needs and priorities of the local community. This is largely to do with their approach to the agendas being set by their training, by government, by professional bodies or employers. However it can also be to do with their personal perception of the way people in the community should be 'managed' by professionals in official positions. And

Figure 11.2 Castleford Queen's Mill with the River Aire in flood © Castleford Heritage Trust

arguably a factor is also their desire to wield power in relationships with amateurs and members of the public.

From the very beginning in 1999, when CHT placed a new clock on the Old Market Hall, we were told by an officer in the local authority Culture Department that it was impossible to celebrate a longstanding tradition to meet under the Old Market Clock, as people had done for generations, as we would not be able to get a road closure. Rather than accept this at face value, we asked Gordon Brown MP, then the Chancellor of the Exchequer, to unveil a plaque and when he said 'yes' the local authority suddenly found we could get a road closure and meet with Gordon under the market clock. We had put up a new clock, with its coat of arms, for the millennium: both the clock, and the tradition of meeting under it, are regarded by our community as an important part of our heritage and our story, but not regarded as legitimate by the local authority. The problem was solved with the intervention of an important personality and this was an important lesson learned.

We remembered this lesson when the Regional Development Agency had a programme of Urban Renaissance and it was said by those in power in the local authority in Wakefield that 'you can't go to Castleford using big words like renaissance as they won't know what you're talking about'. We got Channel 4's *Grand Designs* presenter Kevin McCloud to come to town to do a 'make-over' programme, which became a four-episode television programme called 'Kevin McCloud and the Big Town Plan' (Channel 4 2008). And suddenly we were the exemplar Renaissance Community. Again, when we were not allowed to explore the archaeology in the town centre, Tony Robinson came to make a Channel 4 *Time Team* programme in Castleford. He even called for the return of our town's museum collection which had been taken away to be stored in a warehouse at the opposite side of the Wakefield District (Channel 4 2002).

Castleford people often rightly say that our heritage has been taken away, knocked down, sold off or covered over. Because not only have the authorities seen fit to remove the museum collections, they also sold the Old Market Hall to developers and it has been unused for 25 years except as a walk through to the car park. They knocked down the birthplace of Henry Moore – 'Not worth preserving,' one councillor said. 'Just a run-of-the-mill miner's cottage' (McClarence 2011) – and they covered over the Roman bath-house site even when they were offered the money to preserve it. When the community wanted to put the new museum there to open up the site of the bath-house again, the local authority chose another site at the library.

So walking through town you might be tempted to say what consultants for one of our initial Community Heritage projects remarked on their first visit to Castleford: 'Heritage? What heritage?'. But by the time we had finished putting our case for the unseen and unused, as well as the intangible heritage of our stories, our traditions, our songs and rhymes and dances, our sayings and values, the heritage consultants came back saying, 'It's been like digging up gold all day!' All these experiences could have hindered our progress and dampened our enthusiasm but mostly we did not let it interfere with our work in the community.

Figure 11.3 Castleford Millennium Clock on the Old Market Hall © Castleford Heritage Trust

Figure 11.4 Queen's Mill. Castleford, framed by the iconic curving footbridge designed by Renato Benedetti (McDowell+Benedetti) © Tim Soar

Now to examine some of the issues that face us in the present climate with our current projects.

The Queen's Mill project

The Queen's Mill is on the River Aire in the centre of Castleford, a few hundred metres below its confluence with the River Calder. Previously known as Allinson's Mill, it was the world's largest stone-grinding flour mill, with 20 pairs of grinding stones. The mill venture has huge backing in the local community, with many volunteers offering to get involved and donate to the project. This mill and its story is of interest and delight to people and groups from across the country and beyond. Many make contact by email and visit in person from countries as far away as Australia, New Zealand, North and South America as well as wider UK and Europe.

Yet the Queen's Mill is not a listed building and has no protection or recognition as a heritage site. Even the huge and wonderful waterwheel was largely unknown by national experts, who have said, since discovering it, that it is 'Some wheel! A marvellous piece of Victorian industrial technology, a real gem!'

When examining some of the paperwork left behind I found a letter from the local authority giving the mill owners permission to destroy the waterwheel as they had no interest in it. The former owners, British Waterways, now the Canal and River Trust, commissioned an historic buildings survey to be carried out and presented the CHT with a copy: a short history of the Mill is on the CHT's website (Castleford Heritage Trust 2015). Archaeological work is said to show that grain from the Roman period was found on the site and mills are recorded here throughout the mediaeval period, so it is assumed that milling has taken place at the site for 2,000 years. The mill site is close to the site of the ancient Roman fort and bath-house and mediaeval records show mills on both sides of the river at this point. The present buildings date from between c.1820 and c.1888, replacing a succession of medieval and later structures. The Queen's Mill finally closed in 2010 and the CHT made an offer to buy it in 2012.

Not everyone was pleased when the community decided to buy the mill. We were in competition with developers, who saw the opportunity to try and buy it before the CHT got organised. They could have been one of the early beneficiaries of the Heritage Enterprise grant by turning the site into apartments. Castleford needs development and developers to come to the town, and the river corridor is designated as a Housing Growth Area, so you might think this was a good idea.

However, the members of CHT know the town and its regeneration plans, as we have studied and helped to compile them over the years in various government-funded Urban Renaissance programmes. We need the brownfield sites of former collieries, potteries, chemical works and other industries to be developed, and we need the houses that will be provided, but there is plenty of land for all this around the town without having to use the mill for such purposes.

Castleford is one of the largest towns in the Yorkshire coalfield. The Yorkshire coalfield is the largest in England and Castleford has some of the most severe deprivation in the country. So the Queen's Mill project is more than just about

Figure 11.5 Heritage exhibition in Castleford Old Market Hall © Castleford Heritage Trust

providing a heritage centre for a market town; it is about trying to turn around the fortunes of a whole community, it is about building a better future for our people, for our families.

So currently what are the issues and is it possible that we can succeed?

With the 'Big Society' agenda proposed by the recent Coalition government there has been a government drive for community-led projects, community-owned buildings and community-run services. Hence the establishment of government-funded organisations, such as 'Locality' and the Department for Communities and Local Government (DCLG), providing programmes and grants to help transfer 'ownership' of buildings from local authorities to communities, charities and social enterprises (Locality 2015).

However, when CHT attempted to take action and deliver the dream to make the mill into a community resource, there was no money locally and no one to turn to as the local authority has increasingly massive cuts to its budget each year.

As for other funding for heritage projects, the only grants available are from the Heritage Lottery Fund (HLF) funding organisation, since other government agencies and departments are being asked to take greater and deeper cuts in their budgets.

We might turn to Arts Council England (ACE) or HLF for funding for the development of Queen's Mill as a quality heritage and arts venue. We want to reclaim the unique museum collection with unique Roman artefacts and an Iron-Age chariot burial, to use in an accessible store for educational and research purposes. We want to tell the story of Castleford's famous son, world-renowned sculptor Henry Moore, and exhibit some of his work.

But HLF and ACE have increasingly less money as central government diverts more and more Lottery income to provide services that were previously funded by money from government taxes. The money HLF and ACE give away in grants is clearly to prioritise nationally significant or nationally important projects.

So the funding available for an unlisted flour mill in a de-industrialised backwater like Castleford is not easy to find, even if it is a deprived neighbourhood priority area. The officers at HLF are overworked by the large number of projects coming forward for limited amounts and the competition is fierce so the majority are bound to be turned away disappointed, even if they are worthy causes.

Projects like Queen's Mill have no chance of obtaining the realistic amounts needed for refurbishment and must ask for less than is needed and then try to find other ways of bringing the plans to fruition.

Private businesses and rich benefactors

Towns like Castleford are places that the rich moved out of long ago and many members of the community are struggling to survive in these austere times, when it seems the rich get richer and the poor get poorer. So who is left to give the money we need?

Perhaps locally-based national firms and businesses might be attracted to get behind the Mill project? Perhaps large firms like TEVA (TEVA Pharmaceuticals Ltd), the large international drug developer, manufacturer and distributor which has a huge distribution site located on the edge of town? You might think they could perhaps be persuaded to help out, with all the profits such companies make each year.

But they are globally focussed and, when we were tempted to turn to them, it turned out badly for us, as we ended up subsidising an excellent quality artwork by a very talented local artist, and our small charity was severely out of pocket by the whole venture.

Burberry (the clothing designer and manufacturer) has its largest UK factory in the town, and our community has provided skilled workers for the clothing industry for many generations. We are hoping to be more successful with them, but they are looking for prestigious ventures to support so that their brand name and identity is not tainted and it is far from clear yet that we will succeed. As a consequence, they are being cautious about our requests for support. As success breeds success, we do not want to fail, so we are not pushing too hard. Meanwhile, Burberry are supporting The Hepworth Wakefield (a modern, purpose built art gallery), said to be proving a national success. So will we ever compete with such a prestigious venue?

Even asking for small gifts from local business for a raffle or a tombola is not as easy as might be imagined, as many respond negatively to local groups' fund-raising efforts. They often say 'no' as they have corporate policies to support national charities with higher profiles, such as Macmillan Cancer Care, the Heart Foundation and so on.

So we are left, again, turning to the local authority at Wakefield for small grants and officer help. Ironically, the staff in the Culture Department's museum services may be especially fearful of us because they may feel threatened by our volunteers, keen to get involved as willing unpaid workers. This may be construed as a means of the Council replacing the paid staff and making them redundant, as the local councils' budgets are being drastically cut.

Offering local project services for others

We have been involved in attempting to run limited mainstream services by tendering for management of, for example, a heritage trails site. However, although groups and small charities like our Trust are encouraged to do this, we have to put all the necessary systems in place to comply with their requirements, and it is not easy for us to do this. When we do make the effort, achieve the required standards and try to tender for the work, small organisations like ours are often not trusted: national organisations and agencies are selected rather than give the financial support locally. The bigger organisations get the contracts even when they do not know the local communities that they are commissioned to work with. They do not live in the area, they do not have to face the consequences of getting it wrong, as when the job is done they move on; they just deliver what they must and leave.

An example of this is the Natural England Woods Project (Natural England is the government's adviser on the natural heritage). We delivered a project for this agency close to one of the most deprived estates in the country, even when the National Health Service Primary Care Trust (PCT) (the part of the National Health Service largely responsible for commissioning health care from health service providers) community health officers said we would struggle to get people involved, as they had failed. But we were successful at getting hundreds of people at events and workshops, and woodland or crafts skills training courses were popular. The PCT officers came to see how we were managing to achieve our success and other agencies such as the large district-wide housing organisation, Wakefield District Housing, decided to include us on their local team to help them hit their targets and get the results they needed. But was our success recognised and rewarded, or were we merely acknowledged as a group who could help others meet their targets as and when they needed us?

Covering costs for small charities: effective equal partnership?

A huge problem for small charities like CHT is covering core costs. You can never properly cover core costs from grant applications, so we are constantly struggling to find the money to pay essential bills and staff salaries. For example, a grant from Police and Crime Commissioner Community funds, 'Positive Choices', enabled us to deliver a youth activity project. But the grant was less than £3,000 and only £250 of this could be claimed for project management and administrative costs. We estimated that we did more than double that in paid staff hours, as well as many hours' work by volunteers to deliver the workshops, activities and materials for the projects.

So money available for running the projects often does not reflect the real cost to the communities and local people running them. It seems that large or national organisations and local authorities think it is OK for small community groups to be out of pocket and subsidise things, in a way that would never be seen to be acceptable, and would never happen, for larger agencies such as the Environment Agency.

For instance, when we undertook a Fish Pass project on the river, in partnership with the Environment Agency, they were given large grants and were not expected to report and monitor in ways that are expected of us. They were considered to be such an established national organisation that their procedures were taken for granted, when, in fact, they knew far less than us about efficiently and successfully delivering local projects with the community, and admitted they would have struggled to deliver their results without us.

With the Arts Council, we were asked to deliver some work in the town using a controversial sculpture they had provided. We pointed out that they would not ask anyone else to do it for nothing, so they relented and gave us a small grant (£20,000) to work with local artists and local groups for two years of activities. Following this highly successful project, ACE gave a grant to Yorkshire Sculpture Park (YSP) — a charity and internationally-renowned open-air sculpture park — to work with us for a one-month project to exhibit the work of an internationally known artist, in our gallery. For a project run by YSP with an esteemed and established artist there was a grant of £59,000.

Now trusts like ours are having to compete with large organisations in the District, such as The Hepworth Wakefield, BEAM (a company specialising in art and the public realm) and The Art House (an art community housed in a new building in Wakefield providing space where disabled and able-bodied artists can work alongside each other, meeting rooms, community space and offices), coming for the smaller community grants. And we are all competing for less and less money.

The ACE-funded 'Creative People and Places' arts grant programme, was originally meant to fund local communities (Arts Council England 2015). Yet the bid that was submitted for a project in Wakefield District had no community consultation or involvement in creating the ideas for the project that arts professionals submitted and intended to deliver with communities.

The big issue for us is: how are we going to get real partnerships so that local communities are treated as effective, equal partners. Only then will we really have successful community heritage projects of the highest quality.

So how can the heritage practitioner better engage with the local community?

Our advice would be 'Listen, Listen, Listen!' Be prepared to adapt and work in *equal* partnerships; this will require relinquishing some of the power vested in you by your professional position and expertise.

Partnership working is not easy, but when partnerships work they deliver what those involved desire, and they often create additional benefits for the partners and wider benefits for society. Partnerships can be challenging, sometimes involving fragile coalitions and alliances which need effective negotiators. This requires considerable skill in groups and people management. A high degree of honesty and trust in working together creates better, more successful partnerships.

Partnerships are part of the complexity of modern society. The search for funding and the competitiveness of the public sector economic system demand complex solutions to meet the needs of communities and ensure effective delivery of projects, strategies and policies.

Whilst achieving notable success in using community heritage, the Castleford community has been confronted with issues of 'ownership' and control of heritage objects and projects. We have encountered problems concerning our notion of heritage and heritage use in creating and re-creating sense of place and sense of identity.

These issues are important in any community, but in a socially deprived area such as ours it is vital that they are not overlooked by those working with the community. Partnerships are crucial, especially in towns like ours. We need to work together with professionals and find ways in which the use of heritage can be beneficial in addressing specific local problems and bringing about change. They are the only way most projects and services can be effectively and successfully delivered and the requirements of partners and funders as well as the community can be met.

The stakeholders in community heritage projects are required to look forward to actively create a new future, based on passing on from the past beneficial local heritage. Local people have personal knowledge and experience of their community's heritage and its potential for use, which can deliver tangible and intangible benefits, for themselves and for wider society.

We have found in Castleford that there is still a long way to go in education and training if some heritage practitioners and professionals in the heritage and community sector are to develop the necessary skills and motivation to put people in deprived communities at the forefront of their priorities. A new commitment to working in equal partnership to deliver social benefits by using community heritage is still required. Change is happening, but old ways of working need a fresh outlook: perhaps it is time for a new 'bias to the poor'.

References

Arts Council England 2015. *Creative People and Places Fund* [online]. Available at http://www.artscouncil.org.uk/funding/creative-people-and-places-fund [accessed 2 August 2015]

Castleford Heritage Trust 2015. *Queen's Mill* [online]. Available at http://www.castleford-heritagetrust.org.uk/TheMill.php [accessed 2 August 2015]

Channel 4 2002. *Time Team: A Lost Roman City, Castleford.* Time Team Season 9, episode 10 [online]. Available at http://www.channel4.com/programmes/time-team/episode-guide/series-9/ [accessed 2 August 2015]

Channel 4 2008. *Kevin McCloud and the Big Town Plan* [online]. Available at http://www.channel4.com/programmes/kevin-mccloud-and-the-big-town-plan/ [accessed 2 August 2015]

DCMS 2001. *Libraries, Museums Galleries and Archives for All: Co-operating Across the Sectors to Tackle Social Exclusion.* London: Department for Culture, Media and Sport

Foden, M., Fothergill, S. and Gore, T. 2014. *The State of the Coalfields: Economic and Social Conditions in the Former Mining Communities of England, Scotland and Wales. Sheffield: Centre for Regional Social and Economic Research Sheffield Hallam University* [online]. Available at http://www.shu.ac.uk/research/cresr/sites/shu.ac.uk/files/state-of-the-coalfields.pdf [accessed 20 July 2015]

Locality 2015. *What are Community Assets?* [online]. Available at http://locality.org.uk/our-work/assets/what-are-community-assets/ [accessed 29 August 2015]

McClarence, S. 2011. 'Yorkshire: Carving a new trail through the birthplaces of Barbara Hepworth and Henry Moore'. *The Telegraph* 19 May 2011 [online]. Available at http://www.telegraph.co.uk/travel/artsandculture/8524273/Yorkshire-Carving-a-new-trail-through-the-birthplaces-of-Barbara-Hepworth-and-Henry-Moore.html [accessed 29 August 2015]

12 Dunston Staiths

The reconnection of an industrial monument

Craig Wilson

Introduction

At around 2.30 am on Thursday, 20 November 2003 the Dunston Staiths, a long forgotten industrial structure located on the River Tyne in Gateshead, found itself brought back into the public gaze as a substantial fire ripped through the timber work. By the time the fire had been brought under control, a significant central section of the Staiths had ceased to exist. Machinery once sited atop of the structure had come crashing down into the river and a Scheduled Ancient Monument was in a perilous condition and seemingly helpless against the threats of obsolescence, vandalism and fire. The Staiths were by now in the care of the Tyne and Wear Building Preservation Trust and for that body presented a very atypical project. Considerable liability and vulnerability were clear, with no counter balance of monetary value or future end use. This chapter provides a historical backdrop to the development and life of the Staiths, its decline and challenges and eventual reawakening through a series of partnerships, financial support and, most crucially, re-engagement with the community for whom it was once a source of employment.

From the coalfields of County Durham to an international trade

The phrase 'taking coals to Newcastle' refers to the substantial presence of that carbon resource in the city, but this is only part of the narrative. The city acted as a centre for the coal trade, while the catchment for its extraction encompassed much of the North East of England from rural Northumberland, up the Tyne Valley and deep into County Durham. The wealthy coal barons of the region built fashionable country houses, many now lost to urban encroachment and economic decline, and would share ideas and practice through the North of England Mining and Mechanical Institute, located on Westgate Road in the heart of Newcastle. A significant challenge to the extraction of coal from the North East was the silted River Tyne, requiring the use of keelmen who would take exports in shallow draft boats down to the deeper channels near the mouth of the river. The dredging of the river through the 1860s under the direction of the Tyne Improvement Commission resulted in the eventual removal of the King's Meadow Island from the middle of the river at Dunston. Allied to this, the construction of larger and more powerful iron-clad steamships gradually replacing the reliance on sail transformed the economy of the coal trade from the North East. Particularly on the River Tyne, but also elsewhere throughout the region, was the construction of timber riverside structures known as staiths. Their purpose was to permit the loading of coal onto sea-going colliers located in deep water channels and thus removing the need for the keelmen as middle men. Staiths were located throughout the North East, with six main coaling points on the Tyne alone, including the largest at Tyne Dock, Dunston and West Dunston, Northumberland Dock, Whitehill Point and Derwenthaugh Staiths (Keys and Smith 2000).

Control of the movement of coal by land was undertaken through the use of wagon ways and later by rail. Indeed, the origins and expansion of the rail industry from the region can be closely tied to the extraction of coal, with small independent rail companies developed to facilitate its movement. Following a number of mergers, this activity was overseen by the substantial interests and resources of the North Eastern Railway, which undertook the construction of Dunston Staiths in 1890, with the opening of the north staith to traffic in 1893, with three berths and top-loading gravity chutes.

Imported American pitch pine was used in the construction due to it being robust and flexible; it was set upon piles dropped into the mud and placed at 5.3m centres encompassing 98 frames. Demand resulted in the construction of a south staith, opening in 1903, with the creation of a basin to accommodate further shipping. Coal extracted from the County Durham fields would be brought along purpose-built mineral lines to an elevated approach onto the Staiths at a height appropriate for gravity loading onto the ships. At the height of demand in the late 1930s, 4 million tonnes of coal were being transferred annually through the Staiths. The coal was exported to London, to address the demand for gas lighting, and beyond to Europe and the numerous coaling stations strategically located around the globe to feed the fuel demands of the trade of Empire (ibid.).

The process required a steady supply of labour with railway men, sailors and chute operators, known as teemers and trimmers, whose role was to level out the coal as it was deposited onto the ships. Teemers would be present at the chutes to open the trapdoors on the coal wagons transporting the coal into hoppers and onto the chutes (known as spouts). At high tide, when gravity was not sufficient to achieve the drop, conveyor belts would be utilised. The work presented a high degree of danger, with the teemers being required to climb onto chutes when the coal became stuck and the trimmers being particularly at risk from the next deposit of coal: a number of fatalities were recorded throughout its active use. The Staiths provided not only male employment; the constant traffic of shipping also providing a trade for certain women who could attend to a sailor's needs (ibid.).

Decline and dislocation

The inevitable move from coal as a source of domestic fuel, as a feeder for the energy industry and the development of oil-fired boilers for shipping resulted in

Figure 12.1 Staiths with gap due to fire damage in 2003 © C. Wilson

decline and eventual closure of the Staiths in 1980. It remained as an undefined legacy of the previous industrial age along with the surrounding sites, which were subject to high levels of contamination through previous industrial activity, notably the Redheugh Gas Works and Norwood Coke Works.

In 1990 Gateshead became one of the urban centres of post-industrial northern Britain to stimulate regeneration through the hosting of a National Garden Festival. Previous locations had included Stoke-on-Trent, Liverpool and Glasgow. The site chosen was to follow the southern banks of the River Tyne from close to Lord Armstrong's Swing Bridge, culminating in the main focus in Dunston, with the Staiths acting as both a dramatic backdrop and a visitor experience in its own right. Attracting in excess of an estimated 3 million visitors over its 157 days, the focus was a variety of entertainment rides, environmental and ecological experiences and consideration for the industrial heritage of Tyneside. Various works were undertaken on the structure, including the salvaging of pitch pine piles from the remnants of the less significant south staith, which had been largely dismantled in the 1970s to aid the repair to the north section. Over its entire length damaged and rotten elements were removed, with a new surface decking applied and new staircases to aid the visitor flow. Fire risers were installed and accommodation was made for fire appliances. These works provided full access along its 527m. On conclusion of the Festival, however, the site suffered the same fate as others in the UK: its challenging scale, lack of strategic planning and low demand for the land resulting in mothballing.

In 1997, through the absorption of the North East Industrial Monuments Society, ownership of the Staiths passed to the Tyne and Wear Building Preservation Trust, with the joint designation of Grade II listing (1983) and Scheduled Ancient Monument status (1986). Elsewhere on the banks of the Tyne, regeneration was transforming the urban relationship to the river. At Dunston a partnership between Taylor Wimpey and Hemingway Design saw the land adjacent being redeveloped for housing, with the first properties being put up for sale in 2003. With the site being named Staiths Southbank, the long forgotten monument was now playing a branding role in the redevelopment and, although the interaction between the residents and the structure would be only a visual one, it set in motion the consideration of the Staiths as a site worthy of viewing and a celebration of the industrial heritage of the region. Taylor Wimpey undertook research in to its history and celebrated this in a booklet in support of their residential development that complemented the publication, in 2000, of *Steamers at the Staiths: Colliers of the North-East 1841–1945* by Dick Keys and Ken Smith, which stimulated an understanding of the role of the various staiths in the extraction from the North Eastern coalfields (Taylor Wimpey 2003; Keys and Smith 2000).

In this context, the fire of November 2003 presented a striking realisation that the structure could be lost before its full potential in the context of the surrounding regeneration was realised. The cause of the fire, whether deliberate or the result of careless action by youths known to fish from the structure, was never fully ascertained and no arrests were made. The following years saw frustration at attempts to secure the site from unwarranted access. It was clear to all involved that a comprehensive strategy involving a number of partnerships would be

Figure 12.2 Fire damage to central section of the Staiths © C. Wilson

required if what was believed to be the largest timber structure in Europe was to survive. Further subsequent fires only reinforced this, with one in 2010 resulting in the loss of a section of the top layer exposing the timber structure beneath.

The basis for a new strategy was informed by the writing of a conservation plan in 2006 by Donald Insall Associates. A new approach would, at its first level, define what exactly this industrial monument was, how it was to be interpreted in its redundant state and in which ways that could be integrated into a viable and sustainable solution. There was, singularly, no comparable structure to look to for ideas. No other staiths survived intact – the closest being at Blyth, which had much of its structure reduced to a lowered walkway. The nearest in form were the Victorian seaside piers dotted around the coasts of Britain whose challenges, trauma and loss have been well documented. However, in addition to usually being wrought or cast iron in construction, they are intrinsically still functioning in the role they were originally intended for and, indeed, the main impact on them has been a change in social and cultural tastes and a decline in the domestic holiday market that had first stimulated their construction. Similarly, other major industrial sites that had been conserved were largely able to provide a secure visitor experience that encompassed large elements of the former process or provided space for adaptation and commercial exploitation.

The Staiths could provide no such opportunities and, with the Trust understandably hesitant to enter the market for purchasing coaling ships or, indeed, the coal itself, its redundancy was its most visible activity. In addition, the structure could be viewed with ease from both banks of the river. Its relationship with its hinterland in physical terms and with the community that was once actively employed on and around it presented its most significant and untapped asset. Within its timber beams and struts lay the social history of people and place. The speed of industrial decline within the UK, allied to the sanitisation and neutralising of those former sites, meant that, like many other places one had to look hard along the banks of the Tyne – from the business parks to the luxury office accommodation and cultural experiences – to visually interpret what was once a noisy, dirty and intense waterway busy with workers.

The most remarkable experience of the Staiths when visiting was often one of silence, the lapping of water and presence of wild fowl giving contrast to a site which, in its heyday, would have been in a constant state of activity. It increasingly became clear that the ecological and environmental aspects of the site, through its obsolescence, were significant in themselves. The adjacent Saltmarsh Garden on the River Team had been designated a Site of Nature Conservation Importance in 1985 and more recently there had been the awareness that migratory birds were using the timber elements to roost. Furthermore, the silting up of the basin provided for a diversity of environment, with mud flats at low tide.

A partnership to progress

While the Tyne and Wear Building Preservation Trust were the legal owners and custodians of the structure, they sought to pull stakeholders together in a collective action. This approach was stimulated by a clear desire for the structure to not be viewed in isolation but in context to its surrounding area, in both environmental and social terms. Key to this was Gateshead Metropolitan Borough Council, which owned the approach to the Staiths and would be responsible for the surrounding area. Taylor Wimpey had temporarily taken ownership of the land adjacent, including the Saltmarsh Garden, and on completion of their development a public walkway and Keelman cycle route would be created, running under the spur to the Staiths. English Heritage provided guidance and financial assistance on the philosophical and practical approach to the repair of the damaged elements.

A partnership was formed to take forward an application to the Heritage Lottery Fund (HLF) by Martin Hulse, Director of the Trust, acting as Project Manager with English Heritage, Gateshead Council and Durham Wildlife Trust. Intrinsic to that bid would be the necessity of funding the repairs, increasing access and improving health and safety for potential visitors, along with assessment of the environmental and ecological significance of the site and engagement with the local communities and groups. It was clear that there was a great wealth of social history which should be collated to bring the structure to life. This offered potential to work with local schools to learn from its form and setting, as well as considering how the area of Dunston was once an active industrial community.

In 2012 the Trust commissioned Royal Haskoning DHV to undertake a full structural survey, which credited the durability of the Staiths as being structurally sound. In preparation for the bid the Trust also commissioned Durham Wildlife Services to undertake an Ecological Enhancement and Management Plan of Dunston Staiths and associated land, including the Saltmarsh Garden. Investigation by Durham Bird Club identified a number of species present throughout the winter season, including lapwing, redshank, golden plover, greenshank and sandpiper. In addition bats, otters and seals had been spotted. This information proved significant in terms of scheduling both repair works and public access onto the Staiths while becoming an important asset in relation to the variety of potential visitors to the location.

In parallel with this, Northern Architecture and the North of England Civic Trust were engaged to develop the Activity Plan to encompass both the Staiths and the Saltmarsh Garden with the stated vision that:

> The Staiths and Saltmarsh will be widely recognised and used as a coherent and sustainable heritage site and visitor destination, where local ownership, understanding and pride in the site and a strong volunteer base will support public access, education, surveillance, securing long-term preservation and maintenance.
> Northern Architecture and North of England Civic Trust 2014

All parties involved were clear that one of the challenges to the project would be to sustain the interest and awareness beyond the inevitable fixed term of the funding and activities.

The success of the bid resulted in the award of a grant of £418,900 (49 per cent of the total project cost of £861,482). The additional required funding for this phase was sourced through English Heritage, LEAF, Sir James Knott and Garfield Weston Trusts, while online appeals were made to raise £5,000 in public donations. This would permit the crucial repairs to frames 1–36, a number of which had been subject to fire damage, with Owen Pugh employed as the contractor to undertake the works. On completion this project would enable the realisation of a pedestrian and cycle loop onto the Staiths throughout the summer months. Environmental works were undertaken to remove invasive species and restore the saltmarsh, with bat boxes and an otter holt installed. In terms of dissemination, information panels were created, with leafleting and a dedicated website.

Developing an Activity Plan

With high priority given to engagement, a two-year education plan was created to encompass working with local schools, youth groups and development of an oral history project. Northern Architecture successfully tendered for the Activity Plan and was engaged on the project. The funding crucially also provided for the creation of an Activities Manager and Volunteer Coordinator, appointment in 2014.

The district of Dunston ranks among the 20 per cent most deprived areas within the UK, with income deprivation affecting both young and old and indices on education, skills and training being particularly high. However, the relatively low crime levels and quality of living environment suggested that a sense of stability and community existed within the area. This fed into the Activity Plan in terms of its focus on education and engagement with local community groups.

Keen to identify local engagement in the project, a call was made in the area for anyone interested in the Staiths to attend a meeting which led to the formation of the Friends of the Staiths group. They have taken on the role of key holder with responsibility for opening the gates at set times and monitoring access while helping to plan and deliver events and activities for the open season. They deliver talks on the method of construction and the industrial history of the area. The opening of the Staiths Café Bar at the east end provided for an unofficial 'visitor centre', with the opportunity to display leaflets and posters. Nearly 40 residents from around Dunston, including the café, signed up and quickly organised their group; monthly meetings are held with the support of the Trust.

Interaction with local schools was channelled through the focus on the STEM Education Programme. Northern Architecture had an already developed relationship with Kingsmeadow Community Comprehensive School, which was enthusiastic and engaged as the main education partner. The nature of the site also presented the opportunity to support teachers through both history and geography curriculums. This was developed through a series of workshops and site visits, with sensory and material-based activities. These have been varied and included a drop in, model making, pin-hole cameras and creating shelter structures, which were exhibited on the Staiths during Heritage Doors Open Day. A group from Dunston Drop-in Centre were taken on a birdwatching tour, with guidance from experts from Gateshead Council and Durham Wildlife Trust. A walk was arranged following the course from Tanfield Railway, which originated as a wagon way in 1725 and subsequently fed coal to the Staiths. Kingsmeadow School engineering pupils undertook a project to model the structure using various materials to test stability (Northern Architecture 2015).

The Staiths Stories project sought to engage with young and old through both past experiences and interpretation recording memories by oral history. It was felt important to not assume where the significance in these stories lay, rather that the participants should be encouraged and supported to follow their own lines of enquiry and the focus should be very much community-led. Early taster sessions were developed with the Trust, Gateshead Local History group, Gateshead Libraries Services and Gateshead's Older Peoples Assembly.

Working with project artists Michele Allen and Lindsay Duncanson, volunteers, local residents and visitors were brought together to watch a special screening of archive and recent documentary film of the Staiths. A Pinterest site was created to collate the many period images which came to light through this process of engagement (https://uk.pinterest.com/narchitecture/staiths-archival/).

A visual experience

Parallel to the endeavours of the Trust and various partners and stakeholders, the Dunston Staiths underwent a reawakening in the vibrant cultural and arts scene of Tyneside through a series of events and experiences which brought it back into the public domain. In 2012 the RIBA North East's Forgotten Spaces competition was won by local practice Mawson Kerr, which considered the Staiths as a 'Tyne Tee' driving range, with players lined along the top striking eco-golf balls into the river. In addition to an excellent model, it provided new publicity for the Trust at a significant time.

The Trust acted as a partner to the Jetty Project led by Wolfgang Weileder, artist and Professor of Contemporary Sculpture at Newcastle University, and Simon Guy, Dean of Faculty of Arts and Social Sciences at the University of Lancaster. Defined as 'Art as a catalyst for sustainability' the research project, funded by the Arts and Humanities Research Council, encompassed a three-stage interaction with the Staiths supported by an International Symposium held at the Baltic Centre for Contemporary Art in Gateshead in 2014.

Working in partnership with Gateshead College and Mears Group, the first installation saw the creation of a temporary structure titled 'Cone'. Nine metres in height, it provided a dramatic visual statement to the eastern end of the Staiths and in its form it referenced the bottle kilns once common to the region. Through interaction with the Trust, a number of site visits were arranged for the public to experience the installation up close and view its material form made from recycled waste plastic. The second stage involved consideration of the gap that was created through the fire of 2003. Working with Newcastle University Architecture

Figure 12.3 The Staiths from the landward approach, with new public access 2013 © C. Wilson

Figure 12.4 Interpretation (community, drop in, wildlife, history, café, schools) recognises multiple partners in the project © C. Wilson

students, a scale model was created of the missing section which was then assembled for the public to view. At first this was scaled at 1:10, with the final form measuring 1:4. The third stage considered bridging the gap through computer modelling, reflecting the real debate on the practical and philosophical approach to this challenge.

The AV Festival is a biennial festival of art, film and music held in various locations throughout the North East during March. In 2014 the theme was 'Extraction', which gave focus to the festival on the raw materials that have been hewn from the landscape to stimulate industrial growth and development. With the Staiths being one of the most visible remnants of the extensive coal industry, the structure was chosen to provide a backdrop to one of the most dramatic events of the festival.

Covering the closing weekend, a twice-nightly visual and sound experience was realised by projections onto the north face of the Staiths created by the industrial music group Test Dept. Viewing was arranged by pre-booked boat trips which sailed up river from Newcastle Quayside, with the performance linking back to the Miners' Strike of 1984, an event which retains deep resonance in the region (AVFestival 2014). Further recognition came when the Eat! Newcastle Gateshead festival, held in August 2014 provided the unique experience of dining on the Staiths along a 600m table with a menu inspired by its past, present and future.

Figure 12.5 The Staiths Food Market 2016 © C. Wilson

Conclusion

The next stage is to consider the gap; in this Trust has been encouraged to consider a bold solution. The Staiths stand as a redundant structure, the fire recognised now as a layer of history, its charred edges testament to that ferocious episode. To rebuild the lost elements would be of great expense when a cheaper solution can achieve the same outcome. The desire is to create a safe crossing to relink the eastern and western sections along the top level, permitting a complete circuit to be achieved for visitors. This presents the opportunity to consider how a new engineered structure will interface the timber form, becoming an integral element of design and modern construction value while providing an experience of viewing the remnant elements from above.

In March 2015 the Trust celebrated the completion of works with a launch event held on the Staiths, with attendees from Gateshead Council, the Heritage Lottery Fund and English Heritage among members of the public. Following this, the Staiths opened twice weekly throughout the summer months at a nominal fee, the charge created in consideration of VAT arrangements, with access arranged through the support of the Friends and volunteers group. Visitors have proved to be both local, some once workers there, and from considerably further afield. The opening of the cycle way has brought cyclists onto the top in a rare experience. It is a challenge to the Trust to continue daily management of the structure and for a genuine lasting arrangement; the role of the Friends of the Staiths and schools and youth groups is core and central to its future. Informed, enthusiastic and organised, their sense of custodianship within the community will permit the stories of Dunston to continue. The project was summed up by Martin Hulse, Trust Director:

> At the beginning we couldn't prove that the structure could be kept, we couldn't show interest. Our big surprise was the affection for the Staiths and how quickly people came forward offering their help and support and to organise. To believe that this was something that was worth spending money on, to commit the Staiths to a future.
>
> Hulse 2015

References

AV Festival 2014. *Extraction: Test Dept at Dunston Staiths*. Available at http://www.avfestival.co.uk/programme/2014/events-and-exhibitions/test-dept [accessed 25 August 2015]

Brazendale, A. 1998. *Wickham, Swalwell and Dunston*. Gloucestershire: Tempus

Durham Wildlife Services 2014. 'Ecological enhancement and management plan' [pdf]. Unpublished report

Hulse, M. 2015. Interview with Martin Hulse, Tyne and Wear Building Preservation Trust, June

Keys, D. and Smith, K. 2000. *Steamers at the Staiths: Colliers of the North-East 1841–1945*. Newcastle: Libraries and Information Services

Keys, D. and Smith, K. 2006. *Tales from the Tyne*. Newcastle: Tyne Bridge

Northern Architecture 2015. *Staithes and Saltmarsh Garden Activity* blog. Available at https://staithsandsaltmarsh.wordpress.com/ [accessed 25 August 2015]

Northern Architecture and North of England Civic Trust 2014. 'Dunston Staiths and Saltmarsh Garden Project: executive summary'. Unpublished report

Taylor Wimpey 2003. *The Life of the Dunston Staiths*. London: Taylor Wimpey

The Jetty Project 2015. Available at http://jetty-project.info [accessed 25 August 2015]

Tyne and Wear Building Preservation Trust 2015. *Dunston Staiths*. Available at http://dunstonstaiths.org.uk/ [accessed 25 August 2015]

13 Heritage conservation

The forgotten agenda in Victorian terraced communities

Joanne Harrison

Introduction

Victorian terraced houses are familiar to everyone but are not always a valued contribution to our built heritage. As a result of regeneration programmes, sustainability, and health and safety legislation, these houses have come into focus as a type that needs considerable intervention if twenty-first-century housing needs are to be met.

In many areas, official data shows that the terraced communities are suffering from a variety of problems. Some are related to the fabric of the house, such as poor thermal performance, failure to meet the Decent Homes Standard and a spatial design that is less suited to modern living than more recent builds; and some are related to the community, such as a poor quality environment, social and economic problems, empty homes resulting from oversupply of the type, and failing housing markets (Barraclough *et al.* 2008a; NAO 2007: 10). While this 'official' view of the communities has some validity, the social researcher Allen (2008: 106–8) has argued that there is also a view from the communities of their 'lived' experience, which is often more positive, but attributed less legitimacy because of the residents' working-class backgrounds.

Community engagement is a well-established practice in housing and neighbourhood regeneration programmes, but the inclusion of heritage and conservation issues on the agenda appears to be minimal, and they are rarely a consideration in the formulation of proposals for communities of Victorian terraced housing.

This chapter will review critically how the ideals for participatory engagement by communities in housing renewal have been enacted in practice and how heritage values have figured, if at all, in this process. It considers two distinct types of terraced house. The through terrace has been at the centre of national controversy because of the Housing Market Renewal Initiative (HMRI), which identified tens of thousands as obsolete on the grounds of failing housing markets, while the back-to-back terrace in Leeds is at risk because of high concentrations in the inner city areas which are characterised by social deprivation (Barraclough *et al.* 2008a: 10; Leather and Nevin 2013: 857).

Context in the literature

Muthesius's seminal work *The English Terraced House* covers the history of this housing type, written from an art-historical perspective, discussing typologies, architectural styles, and technological development (Muthesius 1982), while Beresford (1971, 1980, 1988) concentrated on the historical development of the back-to-back in Leeds. More recent publications are concerned with the adaptability of terraced housing, based on the premise that the stock is essentially sound and viable, and therefore has a place in the twenty-first century (Mark Hines Architects 2010; Wilkinson 2006; Yates, 2006).

The *Building Regulations* (HM Government 2010) and *Conservation Principles* by English Heritage (now Historic England) (English Heritage 2008) provide important background information and a policy framework for the issues raised during housing renewal and regeneration in the historic environment, while Lister *et al.* (2007) provide guidance on community engagement and capacity building processes relevant to housing regeneration, and Allen (2008) argues about social class and engagement in the HMRI process. Waterton, Smith and others explain the forms that heritage and community can take, having developed understandings of the ways in which communities may participate in heritage as a process (Little 2007; Perkin 2011; Smith and Waterton 2009; Waterton and Smith 2011). It appears, however, that there is little research specifically into heritage engagement during housing regeneration.

Official publications and those written by experts involved in the HMRI draw heavily on data and statistics, with much less focus on human experience (Audit Office 2004; Cole and Flint 2007; NAO 2007). Case studies are distributed throughout this literature, but are detailed more extensively on stakeholder and media websites (BBC 2013; Liverpool City Council 2007; SAVE 2013b; Edge and WSHG 2011a). A two-volume publication about the back-to-backs in Leeds includes a wealth of research, from historical, housing and demographic analysis to community opinion and costed design proposals (Barraclough *et al.* 2008a, 2008b).

Valuing terraced houses

Many terraced communities have a distinctive sense of place, with a high density, often grid-iron urban grain, an instantly recognisable architectural character with a relatively limited material palette, and standardised plan forms and elevations, yet they encompass a variety of Victorian detailing styles (Muthesius 1982; Power 2007; Wilkinson 2006). Although many of the houses may not have significance and aesthetic value as identified by English Heritage (2008), some will have group value, or historic or communal value which can be measured by the intangible benefits they bring to the resident communities (Wilkinson 2007: 21–2).

Terraced houses are an adaptable form (Power 2007: 7; Wilkinson 2007: 21). There are examples in every town of houses that have had few adaptations since being built, while others have been rearranged internally to provide more open plan living accommodation, combined with adjacent houses to make larger family homes, or even converted to houses for multiple occupation or non-residential use. Research has shown that such houses are regularly re–evaluated and that once unpopular types and neighbourhoods often become sought-after again (Allen 2008: 139–40; Wilkinson 2006: 54). However, some have argued that long-term sustainability remains unlikely to be achieved because of the trend for populations to move from inner city areas to suburbs (Leather and Nevin 2013: 870).

In environmental terms, the performance of many terraced houses is poor, but it is possible to make improvements. Carrig Conservation *et al.* (2004: 4) carried out a study proving a refurbished existing building could perform better environmentally than a hypothetical new building on the same site. The level of intervention has a relationship with loss of historic fabric, and it can be considered that there is a conservation threshold, at which further environmental improvements are detrimental to the building's heritage values (Yates 2006: 10, 17). Arguably, however, even if improvements are limited to this level, it is still a more sustainable solution than demolition and rebuilding because of the embodied energy conserved in the process (Wilkinson 2006: 55; Yates 2006: 18).

There is little heritage protection in place for typical urban or workers' terraced housing. Outside of conservation areas, there are few restrictions on the alterations that owners can make, especially when permitted development rights are exercised. Visually, this has eroded the original character of many neighbourhoods, as is demonstrated in the street in Wakefield, illustrated in Figure 13.1. It includes the birthplace of the sculptor, Barbara Hepworth, but no heritage protection is in place and inappropriate changes have been made to windows. While the Building Regulations do make special allowance for the environmental performance requirements of traditionally constructed houses being renovated, there are no restrictions, so buildings are not protected from inappropriate work which may damage the fabric due to differing building pathologies in traditional and modern construction methods (HM Government 2010: 10). Neighbourhood Plans could be a useful tool in limiting alterations to those which are appropriate to the historic setting, incorporating specific policies and design guidance which, once adopted, would become part of the statutory development plan enabling planning applications to be determined in accordance with their requirements (DCLG 2012: 15, 43–4, 2014a, 2014b).

According to Allen (2008: 178–9), the idea of housing as an investment opportunity is not a factor in the lives of many working-class residents – they are concerned largely with how well it meets their housing needs. However, it is not just the micro-economics of an affordable tenure and running costs that need to be factored into any analysis of their value and suitability, but the macroeconomics of sustaining a community where people want to live, where the houses and environment are pleasant, meet modern living requirements and decency standards, and where there is a healthy economy with business activity and a positive housing market. It has been claimed that, in some regeneration programmes, economic issues have taken precedence over heritage and

Figure 13.1 Inappropriate modifications to fenestration erode the historic character of terraced houses and streetscape, Wakefield © Author

environmental issues, and even over the residents' wishes (*Tonight with Trevor McDonald* 2005a; Wilkinson 2006: 66–7). Lichfield observes the phenomenon found where buildings are declared obsolete not because of their intrinsic defects, but because the site is economically obsolete and its value is higher for redevelopment following demolition (1988: 132).

Engagement

Community engagement and active participation are vital elements of the regeneration process, but their meaning can be interpreted in many ways (Smith and Waterton 2009: 15–16). Little (2007: 2, 4–5) cites definitions concerned with long-term relationships among stakeholders that build knowledge and skills to research and interpret environmental, social and cultural issues with the aim of helping people understand past lives. Engagement can be top-down – where, for example, government sets the agenda for a project it has committed to – or bottom-up – where the need for a project and its approach is determined by the community itself, with guidance from professionals (Perkin 2011: 115, 125). The process is often fraught with tensions related to power imbalances, differing agendas, politics and cultures (Smith and Waterton 2009: 18–19; Perkin 2011: 116). It is claimed that among the barriers to engagement is a tendency for white middle-class communities to be given a higher status (Waterton and Smith 2011: 18, 20–1; Watson and Waterton 2011: 4). Communities outside of this group may be less able to empower themselves and assert their objectives, articulate what they want and interact on the same levels as professionals in heritage matters. There is, therefore, a need to increase the opportunities for participation and build capacities so that stakeholders understand the issues, not just in terms of how they as individuals are affected by a change in housing and local heritage, or the power and influence they can have, but also the wider strategic objectives of a regeneration programme, statutory powers and processes (Cole and Flint 2007: 4, 14).

The key processes characterising successful engagement are, it is suggested: identification of stakeholders; understanding the neighbourhood and the value of heritage to the community; formation of partnerships for collaborative research, outreach and publicity; actively seeking committed participation; and being willing to refine project outcomes in accordance with community responses while remaining within realistic limits (Gadsby and Chidester 2007: 224; Lister *et al.* 2007: 20–3).

This might involve a whole range of activities, such as: neighbourhood audits, profiles or surveys, community walkabouts, stakeholder interviews, collating and sharing data, contacting key people and residents associations in the area, holding public meetings, 'piggy–backing' on established events, and holding events such as community planning weekends, 'Planning for Real' design games and ideas competitions. In addition, capacity building activities could include resources to support the community skill base in the technical aspects of regeneration: help with understanding plans and developing design ideas, accessing research and advice on legal issues, government policy and securing funding. Even general skills such as communication and dealing with community problems, and organisational skills such as running meetings and negotiation will be important for success (Lister *et al.* 2007: 114, 50–4; Power 2007: 14). Negotiation is essential to the process, and communities and heritage experts may need to reach acceptance of each other's authority in arriving at a mutually acceptable conclusion (Gadsby and Chidester 2007: 237–8; Smith and Waterton 2009: 140).

So to what extent is this happening and does the reality of community engagement in the regeneration of Victorian terraced communities meet the best-practice ideals?

Housing Market Renewal Initiative

The programme

The Housing Market Renewal Initiative commenced in 2002 with the opportunity for nine 'pathfinder' areas to produce a prospectus setting out their strategy for renewing neighbourhoods by changing the housing market (Lister *et al.* 2007: 14; NAO 2007: 6). The chosen areas had the most extensive housing market failure, and each consisted of more than one local authority area (Cole and Nevin 2004: 19). They were characterised by a long history of decline with weakened local economies, high levels of deprivation and anti-social behaviour, properties in poor condition, abandoned properties, reducing populations and a lack of variety in the housing stock (NAO 2007: 10). Interventions available to pathfinder programmes included the acquisition of land and property, the clearance of obsolete and surplus housing stock, new build schemes and the renovation and refurbishment of existing properties (NAO 2007: 12).

Although the strategies were put together before 'pathfinders' were allocated funding, and this could include the vision for the area and descriptions of local interventions, the approach to consultation was a key issue (Lister *et al.* 2007: 14). The expectation was that local communities would be engaged, and have a *real opportunity* to identify problems and shape solutions (NAO 2007: 12).

Following acceptance of the prospectuses, government funding was allocated from late 2003. As the programme progressed, it became clear that, in addition to housing supply, there were many factors shaping the market. Local economic performance, demographic trends, interest rates, investor confidence, stock management by local authorities and Registered Social Landlords, community cohesion and metropolitan abandonment all played a part. The condition of the stock and the neighbourhood, it was conceded, might not be a driver so much as a symptom (Cole and Nevin 2004: 28–9; NAO 2007: 24).

Critics of the scheme claimed that the programme was characterised by an obsession with demolition, particularly of Victorian terraced houses (Mark Hines Architects 2010; Leather and Nevin 2013: 870; Wilkinson 2006), and this brought widespread controversy relating to heritage, community engagement, environment, social and economic issues. Cole and Flint (2007: 2) found that the 'pathfinders' were using 'Enquiry by Design' and 'Planning for Real' exercises to increase engagement and to aid understanding of the need to balance the residents' needs with more strategic requirements for the neighbourhoods. However, the NAO (2007: 15) noted that, in the earliest schemes, time was not taken to research and understand the housing markets and heritage significance of the housing stock, while others raised concerns about the motives behind such actions, namely, maximising profits for new build developers (*Tonight with Trevor McDonald* 2005a). There have been a number of high-profile campaigns where local communities who were denied the opportunity to engage in deciding the future of their homes have brought their case to public enquiry (Cole and Flint 2007: 14).

The 'pathfinder' programme ran until 2010, when the (then) coalition government terminated the funding (Wilson 2013: 5). Transition funds were awarded from 2011 to assist with rehousing residents of those areas suffering with blight as a result of incomplete projects (Leather and Nevin 2013: 858).

Welsh Streets, Liverpool

The Welsh Streets in the Princes Park Renewal Area of Liverpool comprise rows of 'two-up two-down' through terraced houses. Their significance can be said to lie in a mix of historic, aesthetic and communal values. The area was designed by a prominent Welsh architect, and built by Welsh artisans who had arrived in Liverpool for work and named the streets after their homeland. One of the streets, Madryn Street, was the childhood home of Ringo Starr, and he also had close connections in adjacent streets, living in the area until he moved to London following success with The Beatles (SAVE 2014; Wilkinson 2006: 22; Edge and WSHG 2011d). Nearby Kelvin Grove is lined with four- and five-bedroomed three-storey terraced houses (Edge and WSHG 2011d).

At the start of the HMRI, about 30 per cent of the houses were owner occupied, the rest owned by Liverpool City Council (LCC) and housing associations (Edge and WSHG 2011a). The condition of the properties varied, with some in excellent condition and 7 per cent classified 'unfit' (Edge and WSHG 2011d). Merseyside Civic Society (2006: 2) described the area as having considerable character, with tree-lined streets, attractive terraced properties and heritage qualities that were an asset to the streetscape. Similarly, the Welsh Streets Homes Group (WSHG) believes the area to be popular, and disputes the finding by officials that the area was suffering from low demand and extensive market failure. In a 2003 survey about housing issues and aspirations, 72 per cent of respondents reported that they were satisfied or very satisfied with their existing home, 50 per cent were satisfied or very satisfied with the housing quality in the area and only 1 per cent felt demolition would improve the area (Mott MacDonald 2003: 2, 11). It was proposed, however, to demolish around 500 Victorian homes, build 370 new homes and run a smaller scale refurbishment programme in selected areas (Liverpool City Council 2007: 16; Edge and WSHG 2011c).

Differing viewpoints emerged among the community, and the proportions of those in favour of refurbishment or demolition also changed with time, which

Figure 13.2 Voelas Street before and after the *Pathfinder* programme © Mark Louden (before) and Nina Edge (after)

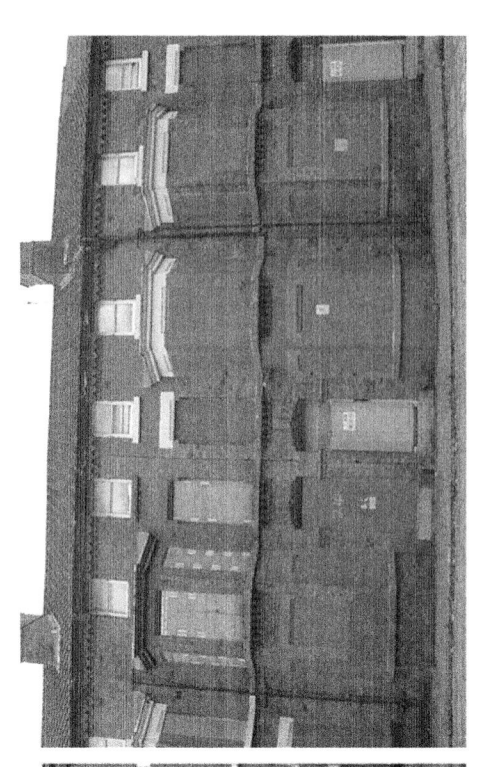
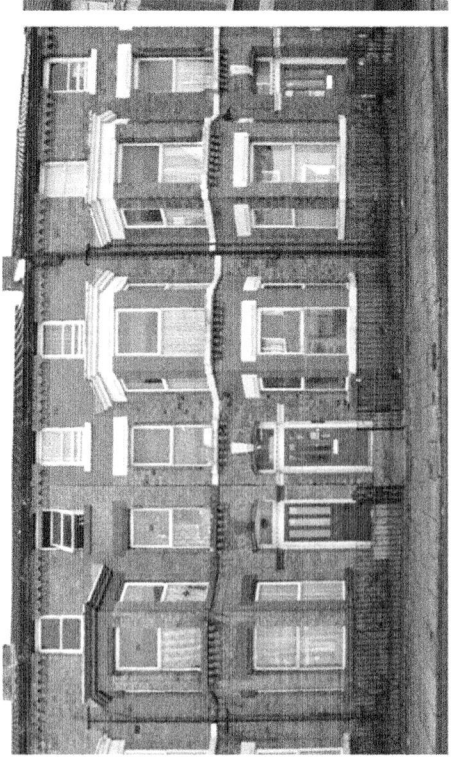

Figure 13.3 Kelvin Grove before and after the *Pathfinder* programme © Mark Louden (before) and Nina Edge (after)

reduced opportunities for cohesion and a unified approach to the regeneration engagement. The residents who were opposed to demolition said they were happy in their homes, liked their community, appreciated the heritage and had environmental or economic concerns about demolition and relocation to new build properties (Allen 2008; 177–9; *Tonight with Trevor McDonald* 2005a; Edge and WSHG 2011b, 2011e). The WSHG believes that repair and refurbishment should be key to revitalising the area and providing sustainable homes, and that new build should be reserved for vacant plots or where houses are beyond repair. They campaigned for this for more than ten years, benefiting from alignment with other local campaign groups and additional campaigns by SAVE Britain's Heritage, the Empty Homes Agency and Merseyside Civic Society (Edge and WSHG 2011d). However, those in favour of demolition felt that the properties were beyond saving, that the area was marred by social problems and that the residents would be happy to move to houses similar to those they had visited as part of the engagement process (*Tonight with Trevor McDonald 2005*a). The *Annual 2006/07* report even notes that an additional three of the Welsh Streets were added to the clearance area upon the wishes of the residents, and that 163 houses had been acquired with the residents' agreement (Liverpool City Council 2007: 6).

What is not clear is how many of the residents supported the heritage agenda independently of their own personal housing, lifestyle and affordability requirements, and how many agreed with demolition because they had not been engaged in exploring refurbishment options.

In 2005, the TV programme *Tonight with Trevor McDonald* showed how refurbishment of a single house on the Welsh Streets could be achieved for under £25,000 and, in 2012, a survey commissioned by SAVE found that it would take £30,000–40,000 to refurbish each property. The properties would be worth at least £35,000 more than their refurbishment cost afterwards, but would still be affordable for those wanting to move into the area. More recently, the cost of refurbishment has been estimated at £51,000–60,000; even this compares favourably to the 2005 cost of demolition and rebuild at between £87,000 and £98,000 (*Tonight with Trevor McDonald* 2005a, 2005b; SAVE 2013a, 2013b. The economic argument for retention therefore seems unanswerable, even disregarding the less easily quantifiable value of the established neighbourhood's heritage.

Wilkinson of SAVE Britain's Heritage expressed serious concerns about the engagement process in his *Pathfinder* report (2006: 51). It was claimed that tender packs were distributed to developers before engagement had commenced, that it is not clear whether residents had any influence on the designation of their streets as a clearance area and that some events from which the proposals emerged were held only during office hours so that many working residents could not participate. Conflicting information about the proposals was given to residents, and street representatives were not actually expected to represent their street to the community steering group. In fact, not all streets had representatives, and those who opposed demolition were not allowed to attend meetings (House of Commons 2005b: 44). Residents were also told that all houses must be cleared in order for any money to be released for area improvements, and this, along with the negative impact of increasingly empty streets as residents voluntarily vacated the area, influenced local opinion in favour of demolition (Edge, personal communication; Wilkinson 2006: 51). The National Audit Office subsequently became involved in the process because of concerns that ballots on demolition versus refurbishment were being re-run until a majority vote was achieved for demolition (Edge, personal communication; House of Commons 2005a, 2005b: 18, 2006).

In 2007, the National Audit Office (NAO) (2007: 38–40) gave detailed information on the form of engagement, which included meetings of community representatives, public meetings and open days, study visits, weekly surgeries, surveys, newsletters and 'vision events', where residents could discuss the proposals and topics presented in an exhibition. It was not until 2007, four years into the consultation process, that a heritage value survey and characterisation assessments were carried out, and a public meeting was held specifically to address the heritage value of the houses. Interestingly, the time of the meeting is noted, perhaps alluding to the fact that its scheduling during office hours would have limited participation. In 2010 it reported the inclusion of a resident to the board, that regular feedback was provided to residents, that training had been given to some residents to increase understanding of regeneration and design standards and that they had actions in place to overcome barriers to participation in the engagement process (Audit Commission 2010: 17). However, there was no indication of the level of influence the residents could have.

Following termination of the 'pathfinder' programme, and upon promise of transition funds (SAVE 2013b), LCC claimed in 2012 that newly revised plans provided by their Registered Provider of social housing, Plus Dane Group, were the result of consultation already undertaken and that residents would be given the opportunity to view and discuss the proposals, enabling them to play a key role in the decision-making process before the scheme was finalised (BBC 2012). Two public consultation events were held, and a website provided information to allow residents to give feedback online (Plus Dane Group 2012). The plans included demolition of 439 Victorian homes and 152 new builds (Hatherley 2013), along with the refurbishment of 32 terraced houses. Following consultation, this was revised to include refurbishment of 37 houses, with retention of a further three properties (BBC 2013). Shortly after completion of the engagement process, the planning application was submitted.

Meanwhile, the WSHG residents engaged a local architect to help them understand the costs of a superior refurbishment to improve the energy efficiency of the existing houses so that they had comparable performance to new build. Having secured agreement from the Housing Minister to run a community design process, they went on to propose a starting point for a consensus solution which would allow both groups of residents – those in favour of demolition and those opposed to it – to write a new brief for Plus Dane Group so that the 152 new build homes could proceed, but more terraced houses could be retained (Edge, personal communication; SAVE 2013a). This was dismissed by LCC and Plus Dane Group and the application was approved, only to be frozen by the Communities Secretary, Eric Pickles. He subsequently called in the application for determination as it was deemed to be of greater than local importance, possibly conflicting with planning policy regarding conservation and enhancement of the historic environment, and had attracted national controversy (DCLG 2013; Plus Dane Group 2013). The public enquiry began in June 2014 and the plans were blocked in January 2015 (BBC 2015; Edge and WSHG 2014). LCC subsequently challenged the ruling, though this was withdrawn in December 2015, and, along with Plus Dane Group, they have developed an alternative plan for the area. Meanwhile, the WSHG secured finance from local companies and organisations experienced in working with Liverpool's communities which would allow more houses to be refurbished. However, the site is now offered to a refurbishment organisation, Place First, who have designated it for private rental after being given exclusive access to develop a proposal. This is rather surprising given that all previous offers to buy and refurbish the area were rejected by LCC, and it brings further uncertainty to residents who are keen to discuss tenure mix and community assets, since another large stakeholder may prove problematic (Edge, personal communication).

The back-to-backs in Leeds

In 2008, an appraisal of the back-to-back terraced housing in Leeds was undertaken following a community-wide consultation on the local development framework (Barraclough *et al.* 2008a; Leeds City Council 2006, 2007; Beattie 2008). The back-to-back terrace is found almost exclusively in West Yorkshire, with the largest concentration, about 19,500, being in Leeds (Barraclough *et al.* 2008a: 10). These houses share a party wall to the back as well as the sides, and have only one external wall. Back-to-backs were the subject of controversy for many years from the 1840s as they were considered to be among the worst type of housing, being associated with overcrowded slum conditions, poor sanitary provision and no through ventilation, which was thought to be the cause of disease. Despite numerous legislative attempts to ban them, their construction was not prohibited until 1909, but a loophole meant that developments with permission prior to the passing of this legislation could still be built. In Leeds, they were the most popular type of housing, forming 71 per cent of the housing stock in 1886, and they continued to be built there until 1937 (Beresford 1971, 1980: 106–7, 1988; Burnett 1986: 70, 74).

Today, 62 per cent of the back-to-back houses are found in inner city areas with mixed communities; where there are high densities, they are characterised by high levels of deprivation and poor environmental conditions. In 2007, 73 per cent of back-to-backs were found to fail the Decent Homes Standard – 50 per cent for reasons of thermal comfort. These factors combined to reduce demand for the property type, while at the same time their value increased to a level where they were becoming beyond the reach of first-time buyers and lower earners (Barraclough *et al.* 2008a: 10, 21–2).

Figure 13.4 The back-to-back terraced houses in Leeds have a distinctive and repetitive form, with most having a basement, two floors and attic accommodation © Author

The aim of the 2008 study was to research the issues affecting the back-to-back houses and communities so that appropriate intervention packages could be proposed, and use of the properties could continue into the twenty-first century. Professionals from a number of agencies were appointed to guide the research and strategy development, including representatives of various Leeds City Council departments, housing-related organisations, the research and strategy developers Re'new and English Heritage (ibid.: 4, 18).

Despite analysing demographics, housing condition, ability to comply with regulations and long-term sustainability, only passing reference was made to the cultural value of the built heritage and its intangible value to the community (ibid.: 18; Barraclough *et al.* 2008b). The community engagement process was comprehensive in many aspects, but the only references to heritage were a note that a separate pilot historic area assessment had been undertaken for one area of the city, and two questions forming part of broadly themed interviews. There were 130 non-back-to-back resident respondents to the quantitative interviews and just 12 back-to-back resident respondents, with a further 11 back-to-back owners interviewed qualitatively.

One question dealt with opinions specifically on back-to-back houses, with 75 per cent agreeing that the heritage value of this housing was important to the image and identity of Leeds and only 12 per cent disagreeing. The other was concerned with how back-to-back housing impacted on the overall impression of Leeds as a twenty-first-century city. In responses, 51 per cent of comments were positive, with 25 per cent of people citing the heritage and history of Leeds, and 45 per cent were negative (Barraclough *et al.* 2008b: 106, 108). The responses figures suggest that back-to-backs do have a perceived heritage value, but that the opinions of residents and owners were not valued as highly as those of professionals or non-residents. It could be speculated that this is evidence of Allen's findings about the reduced legitimacy of the opinions of working-class residents (Allen 2008: 106–8).

The *Understanding Place* series by English Heritage (2010, 2011, 2012) details a methodology for analysing historic areas using a wide range of tools and resources, in order that a place, whether or not it is designated, can be suitably managed, while *Conservation Principles* outlines a method for assessing the significance of a building or place before setting out guidelines for managing change (English Heritage 2008). These rigorous processes exist to ensure that identified heritage values and significance are not eroded and that change takes suitable account of authenticity and integrity, sustainability and reversibility, through evaluation of options (using a heritage impact assessment, for example) and appropriate mitigation methods (English Heritage 2008: 45–7). The National Planning Policy Framework identifies similar heritage protection issues and processes as part of a wider remit covering the government's planning policies for sustainable development (DCLG 2012: 6, 16, 30–2). Unfortunately, in the case of the Leeds back-to-backs, heritage value was not sufficiently recognised to signify that such processes should be undertaken. There was no appraisal of the significance of the houses as a heritage asset, although it was arguably merited given their distribution and historical and cultural associations.

Proposals were subsequently put forward, divided into three levels of intervention. Minimal intervention applied to over 12,000 properties, which were not located in high concentrations, and were concerned only with improvements to meet the Decent Homes Standard. Clearly, here, the problem is identified not with the building form and fabric so much as the social problems associated with high-density communities (Barraclough *et al.* 2008a: 32).

Medium-level interventions were proposed for over 6,000 properties in inner city and inner suburban areas, with the aim of achieving the Decent Homes Standard and improving the appeal of back-to-backs through a range of 'facelift' work to property exteriors, along with general neighbourhood improvements (ibid.: 32–3). With the exception of replacement high-performance timber sash windows, there was no indication that any of the works would be completed using traditional materials and construction techniques (Barraclough *et al.* 2008b: 50–1, 142–55).

Major-level interventions were proposed for just over 1,000 houses, with around 500 being demolished and replaced with new housing, 400 having major internal and external remodelling and the remainder being knocked through to the house behind or alongside or adapted into flats. There was no indication of the extent to which original fabric still existed, or justification for its removal, and intervention options therefore involved the removal of historical elements such as external windows and doors, rainwater goods, internal walls, doors and joinery, the staircase and balustrading, fireplaces, plaster and parts of the floor (Barraclough *et al.* 2008a: 33–6, 2008b: 53–69, 142–55). Ironically, in some remodelling cases the proposals did not even address the regulatory issues that had originally been identified as being problematic. This approach to analysis and design contrasts sharply with all the guidance on planning for historic areas that ought to be taken into account for regeneration in historic neighbourhoods. Neither the distinctive character of the neighbourhood nor the views of local residents on its heritage were integrated in the approach.

The forgotten agenda

It is clear from these two case studies that heritage issues were not a significant factor in the strategic objectives or 'official' agenda in the regeneration of these Victorian terraced communities.

In the Welsh Streets, the programme failed to engage residents in exploring the significance of their built heritage and it was the independent activities of a residents' group, supported by a variety of organisations and the media, that brought the heritage agenda to the fore. The input of heritage professionals appointed through the programme is notable by its absence. Although there is evidence for extensive community engagement, little related to heritage and there are many claims that the community had limited opportunity to influence proposals where this was in opposition to demolition.

In Leeds, there was a detailed investigation of stakeholder opinion on back-to-backs, but nothing that specifically addressed the heritage and conservation agenda. Consequently, the options focused on achieving renovated and remodelled houses that were stripped internally of much of their heritage assets. Although English Heritage had input in the process, there is no evidence to show their official view on the significance of the back-to-back terraces. The West Yorkshire Archaeological Advisory Service, however, has noted in its research agenda that one priority is the need for more research on the architectural character of workers' housing and it is therefore important that the evidence is not destroyed (Giles 2013: 116).

So what procedures should be put in place to ensure that community engagement processes include heritage on the agenda?

Legislation and compliance frameworks could be developed with input from professional bodies, to establish audited processes where heritage and conservation studies form a mandatory element of regeneration programmes. This could be linked with existing planning and development frameworks and possibly work on the assumption that an area's heritage assets and its communities' perceptions of their value should be regarded as significant until proven otherwise. Only when desk-based research and community engagement show a particular historic environment aspect is not significant or obsolete should the requirement for further research, capacity building and community engagement diminish. In addition, a requirement for project teams to include members of the relevant heritage professional bodies would ensure that heritage and conservation matters were being guided by those with appropriate knowledge and experience.

A wide range of tools and techniques can be implemented to engage communities in exploring their heritage and building their capacity in relation to heritage ideas, technical considerations, policies and processes. Exploration and recording of memories in the community's living history, or of anecdotes passed down through the generations, could identify a range of intangible values that not only reinforce the value of a place for the individual, but also the wider community as shared memories are meshed together forming a collective memory and value. Tools could range from informal 'coffee and chat' sessions in established community groups 'to video diaries of individuals or groups reflecting on the past' and formal interviews, the latter perhaps being a second stage of the consultation process to develop thoughts and ideas, and help people assimilate these with value.

Tangible values may be something that Victorian terraced communities are more aware of and are able to identify with more readily. Workshops could operate across the strands of capacity building, understanding communal value and regeneration aspirations. For example, sessions focused on building form, fabric and plot size in terms of space, light and construction quality, architectural character, maintenance, stigma and fashion could encourage people to think about the advantages and disadvantages of their homes, how they live and how well terraced houses do or could meet their needs. Linking this with activities that explore the architecture and construction of traditional houses, including their wider historical contexts (such as socio-economic, political and technological factors) could foster an interest to protect the built heritage. A newfound understanding about why the houses were designed and built as they were could encourage small changes in behaviour or modification choices to improve their suitability as modern homes. Further research will be testing these approaches in practice.

Refurbishment exploration workshops are where professionals and communities can brainstorm, draw, model and develop ideas for retaining and refurbishing the historic fabric. Supplemented with visits to refurbished properties demonstrating the differing levels of intervention that are possible, and improvements and adaptations that could meet a variety of twenty-first-century housing needs, these would allow communities to develop a sense of ownership over proposals which reflect both their own values and needs, and those identified by policy and other regeneration initiatives.

With encouragement from professionals able to understand and promote the heritage agenda in their engagement with residents of terraced communities, and with formal measures in place to protect communities from processes with little opportunity for empowerment and decision-making, the houses could continue to be enjoyed. For their present and future communities there could be real benefit from the historical, social, economic and environmental advantages that conservation-led regeneration brings.

Acknowledgements

I am grateful to Nina Edge for her assistance in helping me to understand the sequence of events in the Welsh Streets consultation, and for alerting me to information and evidence that is not widely known. Thanks also go to her and to Mark Loudon for permission to use their photographs of the Welsh Streets houses; to Gill Chitty and the editing team for their comments on my drafts; and, finally, to my husband and children for allowing me to spend so much time on this research.

References

Allen, C. (2008) *Housing Market Renewal and Social Class*. London: Routledge
Audit Commission (2004) *Market Renewal: New Heartlands Pathfinder*. London: Audit Commission
Audit Commission (2010) *HMR Performance Review 2009. New Heartlands, Merseyside*. London: Audit Commission
Barraclough, J., Horner, D. and Jones, H. (2008a) *Addressing the Challenge of the Back-to-Backs in Leeds. Volume 1: Strategy*. Leeds: Renew Leeds
Barraclough, J., Horner, D. and Jones, H. (2008b). *Addressing the Challenge of the Back-to-Backs in Leeds. Volume 2: Background Research*. Leeds: Renew Leeds
BBC 2012. *Liverpool Welsh Streets Residents in New Homes Consultation* [online]. Available at http://www.bbc.co.uk/news/uk-england-merseyside-19368577 [accessed 23 March 2014]
BBC 2013. *Residents Back Plans for Liverpool's Welsh Streets* [online]. Available at http://www.bbc.co.uk/news/uk-england-merseyside-21016791 [accessed 23 March 2015]

BBC 2015. *Appeal to Welsh Streets Decision to Preserve 'Beatles Heritage'* [online]. Available at http://www.bbc.co.uk/news/uk-england-merseyside-31657716 [accessed 23 March 2015]

Beattie, A. 2008. *A Strategy for Improving Leeds Private Sector Housing*. Leeds: Leeds City Council

Beresford, M. 1971. 'The back-to-back house in Leeds, 1787–1937'. In S.D. Chapman (ed.), *The History of Working-Class Housing: A Symposium*. Newton Abbot: David and Charles, 93–132

Beresford, M. 1980. 'The face of Leeds, 1780–1914'. In D. Fraser (ed.), *A History of Modern Leeds*. Manchester: Manchester University Press, 72–112

Beresford, M. 1988. *East End, West End: The Face of Leeds During Urbanisation 1684–1842*. Leeds: The Thoresby Society

Burnett, J. 1986. *A Social History of Housing, 1815–1985*. London and New York: Methuen

Carrig Conservation, McGrath Environmental Consultants, James P. McGrath & Associates and Murray O'Laoire Architects 2004. *Built to Last: The Sustainable Re-use of Buildings*. Dublin: Heritage Council

Cole, I. and Flint, J. 2007. *Demolition, Relocation and Affordable Housing: Lessons from the Housing Market Renewal Pathfinders*. Coventry: Chartered Institute of Housing

Cole, I. and Nevin, B. 2004. *The Road to Renewal: The Early Development of the Housing Market Renewal Programme in England*, York: Joseph Rowntree Foundation

DCLG 2012. *National Planning Policy Framework*. London: DCLG

DCLG 2013. *Ministers to Decide on Welsh Streets Demolition Proposals* [online]. Available at https://www.gov.uk/government/news/ministers-to-decide-on-welsh-streets-demolition-proposals [accessed 30 April 2014]

DCLG 2014a. *Planning Practice Guidance. Local Plans* [online]. Available at http://planningguidance.communities.gov.uk/blog/guidance/local-plans/ [accessed 29 July 2015]

DCLG 2014b. *Planning Practice Guidance. What is Neighbourhood Planning?* [online]. Available at http://planningguidance.communities.gov.uk/blog/guidance/neighbourhood-planning/ [accessed 29 July 2015]

Edge, N. and WSHG (Welsh Streets Home Group) 2011a. *A Quick History* [online]. Available at http://www.welshstreets.co.uk/?page_id=97 [accessed 23 March 2014]

Edge, N. and WSHG 2011b. *Divided Community* [online]. Available at http://vimeo.com/25186237 [accessed 25 June 2014]

Edge, N. and WSHG 2011c. *Housing market renewal in a nutshell.* [online] Available at http://www.welshstreets.co.uk/?page_id=99 [Accessed 23rd March 2014]

Edge, N. and WSHG 2011d. *Once Upon a Time...* [online]. Available at http://www.welshstreets.co.uk/?page_id=6 [accessed 23 March 2014]

Edge, N. and WSHG 2011e. *The Welsh Streets and 9 Madryn Street: Assorted News Items* [online]. Available at http://vimeo.com/25189963 [accessed 25 June 2014]

Edge, N. and WSHG 2014. *Welsh Streets Public Inquiry Opens: We Can Work It Out?* [online]. Available at http://www.welshstreets.co.uk/?p=1796 [accessed 19 March 2015]

Edge, N. and WSHG 2015. *Are We Appealing or Not?* [online]. Available at http://www.welshstreets.co.uk/?p=1974 [accessed 9 June 2016]

English Heritage 2008. *Conservation Principles: Policies and Guidance for The Sustainable Management of the Historic Environment*. London: English Heritage

English Heritage 2010. *Understanding Place. Historic Area Assessments: Principles and Practice*. Swindon: English Heritage

English Heritage 2011. *Understanding Place: Conservation Area Designation, Appraisal and Management.* Swindon: English Heritage

English Heritage 2012. *Understanding Place. Historic Area Assessments in a Planning and Development Context.* Swindon: English Heritage

Gadsby, D. and Chidester, R. 2007. 'Heritage in Hampden: a participatory research design for public archaeology in a working-class neighbourhood, Baltimore, Maryland'. In B.J. Little and P.A. Shackel (eds), 2007. *Archaeology as a Tool of Civic Engagement.* Lanham, MD, and Plymouth: Altamira Press, 224–38

Giles, C. 2013. *Research Agenda. Historic Buildings in West Yorkshire (Medieval and Post–medieval to 1914).* Wakefield: West Yorkshire Archaeological Advisory Service

Hatherley, O. 2013. *Liverpool's Rotting, Shocking 'Housing Renewal': How Did It Come to This?* [online]. Available at https://www.theguardian.com/commentisfree/2013/mar/27/liverpool-rotting-housing-renewal-pathfinder [accessed 23 March 2014]

HM Government 2010. *The Building Regulations 2010. Approved Document L1B. Conservation of Fuel and Power. 2011.* London: Stationery Office

House of Commons ODPM: Housing, Planning, Local Government and the Regions Committee 2005a. *Empty Homes and Low-demand Pathfinders Written Evidence. House of Commons Papers 2004–05 295-II.* London: Stationery Office

House of Commons ODPM: Housing, Planning, Local Government and the Regions Committee 2005b. *Local Government Consultation. [Vol. 2]: Written Evidence. House of Commons Papers 2004–05 316-II.* London: Stationery Office

House of Commons ODPM: Housing, Planning, Local Government and the Regions Committee 2006. *Affordability and the Supply of Housing Session 2005–06 Volume II: Oral and Written Evidence* [pdf]. Available at http://www.publications.parliament.uk/pa/cm200506/cmselect/cmodpm/703/703ii.pdf [accessed 9 June 2016]

Leather, P. and Nevin, B. 2013. 'The housing market renewal programme: origins, outcomes and the effectiveness of public policy interventions in a volatile market'. *Urban Studies* 50(5), 856–75

Leeds City Council 2006. *East and South East Leeds Area Action Plan. Leeds Local Development Framework. Development Plan Document. Alternative Options – Looking to the Future.* Leeds: Leeds City Council

Leeds City Council 2007. *East and South East Leeds Area Action Plan. Leeds Local Development Framework. Development Plan Document. Preferred Options – The Future Emerges.* Leeds: Leeds City Council

Lichfield, N. 1988. *Economics in Urban Conservation.* Cambridge: Cambridge University Press in association with Jerusalem Institute for Israel Studies

Lister, S., Perry, J. and Thornley, M. 2007. *Community Engagement in Housing Market Renewal.* Coventry: Chartered Institute of Housing

Little, B. 2007. 'Archaeology and civic engagement'. In B.J. Little and P.A. Shackel (eds), *Archaeology as a Tool of Civic Engagement.* Lanham, MD, and Plymouth: Altamira Press, 1–10

Liverpool City Council 2005a. *Declaration of Princes Park Renewal Area HSG/9* [pdf]. Available at http://councillors.liverpool.gov.uk/documents/s13097/HSG.9%20-%20Declaration%20of%20Princes%20Park%20Renewal%20Area.pdf [accessed 9 June 2016]

Liverpool City Council 2005b. *Declaration of Princes Park Renewal Area HSG/9: Appendix.* [pdf]. Available at http://councillors.liverpool.gov.uk/documents/s12968/HSG.9%20-%20Declaration%20of%20Princes%20Park%20Renewal%20Area%20-%20Appendix.pdf [accessed 9 June 2016]

Liverpool City Council 2007. *Housing Market Renewal in Liverpool: Annual 2006/07.* Liverpool: Liverpool City Council

Mark Hines Architects 2010. *Reviving Britain's Terraces: Life After Pathfinder.* London: SAVE Britain's Heritage

Merseyside Civic Society 2006. *Protecting, Preserving and Making Accessible Our Nation's Heritage* [pdf]. Available at http://www.liv.ac.uk/mcs/lfs/consultations/dcmssub0603.pdf [accessed 30 April 2014]

Mott MacDonald. 2003. *The Welsh Streets Neighbourhood Plan Survey* [pdf]. Available at http://www.welshstreets.co.uk/wp-content/uploads/2011/04/Welsh-Sts-hilsvy-report.Chrome.pdf [accessed 9 June 2016]

Muthesius, S. 1982. *The English Terraced House.* New Haven: Yale University Press

NAO (National Audit Office) 2007. *Department for Communities and Local Governmen:. Housing Market Renewal.* London: Stationery Office

Perkin, C. 2011. 'Beyond the rhetoric: negotiating the politics and realising the potential of community-driven heritage engagement'. In E. Waterton and S. Watson (eds), *Heritage and Community Engagement: Collaboration or Contestation?* London: Routledge, 115–30

Plus Dane Group 2012. *Views Sought on the Welsh Streets* [online]. Available at http://www.plusdane.co.uk/views-sought-on-the-welsh-streets/ [accessed 30 April 2014]

Plus Dane Group 2013. *Letter from Plus Dane Group to Eric Pickles, 29.07.2013* [pdf]. Available at https://www.whatdotheyknow.com/request/190020/response/473490/attach/3/140122%20EIR%20Liverpool%20Welsh%20Streets%20Document1.pdf [accessed 19 March 2015]

Power, A. 2007. *Communities and Demolition: Findings from a Workshop at Trafford Hall, the National Communities Resource Centre.* London: LSE Housing, Case

SAVE 2013a. Press release 22 July [online]. Available at http://www.savebritainsheritage.org/docs/articles/Welsh%20Streets%20press%20release%2022ndJuly.pdf [accessed 30 April 2014]

SAVE 2013b. Press release 25 July [online]. Available at http://www.savebritainsheritage.org/docs/articles/Press%20release%2025th%20July%20Pickles_Welsh%20Streets.pdf [accessed 30 April 2014]

SAVE 2014. Press release 2nd June [online]. Available at http://www.savebritainsheritage.org/docs/articles/PRESS%20RELEASE%202%20JUNE%202014.pdf [accessed 12 June 2014]

Smith, L. and Waterton, E. 2009. *Heritage, Communities and Archaeology.* London: Duckworth

Tonight with Trevor McDonald. New Homes or Old. Part 1 2005a. [TV programme recording] ITV, 16 May 7.30 pm. Available at http://vimeo.com/25137004 [accessed 26 May 2014]

Tonight with Trevor McDonald. New Homes or Old. Part 2. 2005b. [TV programme recording] ITV, 20 May 7.30 pm. Available at http://vimeo.com/25144795 [accessed 26 May 2014]

Waterton, E. and Smith, L. 2011. 'The recognition and misrecognition of community heritage'. In S. Watson and E. Waterton (eds), *Heritage and Community Engagement: Collaboration or Contestation?* London: Routledge, 12–23

Watson, S. and Waterton, E. 2011. 'Heritage and community engagement: finding a new agenda'. In S. Watson and E. Waterton (eds), *Heritage and Community Engagement: Collaboration or Contestation?* London: Routledge, 1–11

Wilkinson, A. 2006. *Pathfinder.* London: SAVE Britain's Heritage

Wilkinson, A. 2007. 'Why save empty homes and semi abandoned streets?'. In A. Power (ed.), *Communities and Demolition: Findings from a Workshop at Trafford Hall, the National Communities Resource Centre*. London: LSE Housing, CASE, 21–3

Wilson, W. 2013. *Housing Market Renewal Pathfinders* [pdf]. Available at http://research-briefings.parliament.uk/ResearchBriefing/Summary/SN05953 [accessed 30 April 2014]

Yates, T. 2006. *Sustainable Refurbishment of Victorian Housing: Guidance, Assessment Method and Case Studies*. Bracknell: IHS BRE Press

14 Linking people and heritage

Lessons from community engagement initiatives in India

Krupa Rajangam

Introduction

This chapter describes two community engagement initiatives in India and the learning gathered from them. Both are personal initiatives and stem from an interest to address the questions: 'What do people value?' and 'How does one engage communities in a meaningful dialogue on heritage and its conservation?'. In turn these two questions arise out of what I call the 'conservation conundrum'. This is a dilemma that I face as a practitioner – a heritage expert with over a decade of conservation experience which involves local communities.

The dilemma touches on two themes. First is the prevalent conservation paradigm in India for protected cultural sites. This translates as either 'preservation' – that is, removing all traces of people, habitation and settlement from the site; or 'abandonment' – that is, leaving the site open to encroachment and gradual disappearance (Fritz and Michell 2012). The second theme arises from popular heritage literature in the country. Typically, the voices of heritage experts, enthusiasts and NGOs regret the lack of interest in cultural heritage and its conservation among the general public. They argue for the need to educate people about the history and significance of cultural heritage sites. More recent articles also tend to conclude that, from a sustainability perspective, it is essential that local people be involved in heritage conservation.

However, in general – and particularly in the case of protected sites in India – these local people are very often the same people who were either dislocated in the name of heritage 'preservation' or who 'need to be educated about heritage' in the first place. Thus, in some sense, practice becomes contradictory. Practitioners and the heritage realm are apparently supposed to learn from and educate people at the same time – that is, both understand what people value and simultaneously educate them on what to value and how. I have faced such contradictions in practice on numerous occasions. It raises certain questions, as follows.

First, why *is* it important that conservation practice and the heritage discipline engage with people, including resident communities? Second, if conservation needs to engage with people, why does the preservationist approach persist in India, especially since it seems to marginalise resident communities (i.e. people living in and around heritage sites)? Third, is the issue one of not knowing how to meaningfully engage people or resident communities – is it an operational issue? Finally, what are the positions of resident communities about the cultural heritage site in their midst that has impacted or will impact their lives at some point? Do they value the cultural heritage site? How do they value it? If not what do they value?

These questions were the starting point for my research interest in the role of communities in conservation. A review of disciplinary literature gave some pointers to address the first two (Harvey 2001; Gentry 2014). Although I was unable to locate such studies based on Indian cultural sites, anthropological literature from other cultural and geographical contexts provided some answers to the last question (Bunten 2008; Collins 2008). However, there appears to be a gap in literature that addresses the third question. For example, while Ashley *et al.* (2015: 1) note the positive development from a 'static' to 'dynamic integrated' conservation approach which necessarily engages stakeholders or local communities, in their study of Sudan, they consider that the question of how this might be done 'is less established'.

For this chapter, the focus is therefore on the third question: how can local people and resident communities be engaged meaningfully? I present a summary of two relevant case studies and discuss the challenges in engaging people in a dialogue on heritage and its conservation based on these two initiatives. This is with the aim to begin addressing the issues of how to engage with people but not to seek universal solutions to address the 'how to...'. Rather, it is to indicate the need for a nuanced and contextual understanding of cultural heritage places and the people or local communities who are entangled in them.

The initiatives: *Nakshay* and *Neighbourhood Diaries*

Cultural heritage as a concept, it has been suggested, relates more to people and place than time (Thakur 2012). However, in India, time – particularly its linear expression as age value – continues to define cultural heritage sites or at least 'monuments' and 'protected sites'.[1] Furthermore, the prevalent view of cultural heritage conservation in official circles continues to be that of an expert-driven scientific and technical process, especially in the case of 'protected sites'. While I concur with the view that it is a social process and ought to engage with people, the dominant or 'authorised' view commonly colours the perceptions of cultural heritage and conservation among such groups and individuals. Hence, in an effort to shift the focus onto place and people, we conceptualised two community engagement initiatives: *Nakshay*, where communities map their heritage, and *Neighbourhood Diaries*, which is a collection of stories and histories of neighbourhoods.[2]

Both projects seek to understand people's views on heritage and socio-cultural significance but in different ways. In *Nakshay* our role is more active. We seek different groups, engage with them over a series of sessions and attempt to understand the things, people and places they value the most. That is, we seek to understand their views on cultural heritage without imposing our 'expert' views on them. In *Neighbourhood Diaries* we document and present Bangalore's historic neighbourhoods as short films. Each stand-alone film is a (resident) narrative based on a local socio-cultural space. It incorporates both anecdotes and factual research. In this initiative our role is comparatively passive. We observe and record individual and/or group interactions within local spaces, particularly those that are transforming rapidly. In the process we may catalyse the group to consider that space as cultural heritage and/or proactively manage it. Both are ongoing initiatives; we commenced *Nakshay* in 2010 and *Neighbourhood Diaries* in 2011. However, individual engagements with neighbourhoods and groups are sporadic as both are self-funded and voluntary, owing to the complications involved in seeking funding and employing a team.

Nakshay: community-led culture mapping (see www.nakshay.saythu.com)

First, we conceptualised a personal initiative, *Nakshay* – which translates into 'map' in Hindi – as a community-led, culture mapping project, where communities would both identify and map 'cultural heritage' as defined above. It is an outcome of over a decade of experience in working with different conservation organisations in India as 'heritage experts'. We had observed a lacuna in the system in terms of community engagement, which we attributed to the prevalent expert-led, monument-centric approach. Second, we were influenced by the sustainability concept: a premise that a successful, or sustainable, heritage conservation management system would be one that has developed from within the community. Consequently, we considered that the first step towards realising such a system would be to have communities identify their cultural heritage. Finally, *Nakshay* is also an effort to understand the role of communities in heritage identification and conservation.

It aims to create community awareness of local cultural heritage or socio-cultural hubs. We hoped that the process would instil a sense of pride, foster group identity and maybe result in groups taking a lead in conserving their heritage. Its objectives are twofold: first, to engage with different groups to identify and map places, people, things they consider significant; and second, based on the sessions, to develop guidelines towards engaging different groups in heritage identification.

In keeping with the aim and objectives, we considered the following principles while developing our methodology. First, we would attempt to understand what a group considers heritage or culturally significant through a 'bottom-up' rather than 'top-down' approach. That is, we would neither be prescriptive nor paternalistic. For example, we would not ask individuals or groups if they consider xyx heritage or even what they consider heritage. Rather, we would explore what they consider culturally significant, based on its relevance in their lives. Second, we would not limit our engagement to one-off sessions but interact repeatedly with particular groups. We anticipated that this would ensure a richer understanding. Third, in keeping with our definition of community we would work with different groups, like children, adults, resident associations, to understand multiple viewpoints. Finally, we would make the whole process publicly available by documenting the sessions both as video and text.

Our first *Nakshay* exercise was with a high school class, followed by one with college students. Based on an initial round of discussions with students and faculty, we proposed four sessions in both locations. First, an orientation, followed by a group discussion and interaction, then a deeper individual engagement and, finally, either one to one in the field or presentation to their peers. However, due to academic pressures we had to limit ourselves to three sessions at both locations.

At the school we interacted with 41 students of a tenth-grade class, aged between 14 and 15, in the presence of their history teacher. Our choice of school was deliberate – a state government-run establishment, which was founded in 1905 by the British as the first English medium school in the state. It is housed in a historic colonial-period structure and located in the heart of Bangalore's old fort area. The site is practically surrounded by nationally and regionally protected built heritage. We wondered if the presence of 'protected sites' in the vicinity of the school would influence student choices.

The first session was exploratory and engaged all the students of the class in an effort to build a rapport with the group. We commenced the session by asking the class to tell us something unique about the area around their school, home or en route. In the second session we discussed the object, place, thing or person each student had noted as 'unique' – that is, with socio-cultural significance – to delve into their choices. For example, two students had both noted Tipu's fort as 'something unique'. For the first student, this was because of its association with Tipu Sultan, a significant eighteenth-century ruler. The second student, however, thought the place significant because it was a fort, a defensive structure. We were thus able to use this example to talk about how the same place or thing could have multiple meanings and associations. We then asked the class to rank each other's choices and led this exercise on to a conversation on why some choices were ranked higher than others. At the end of the session we asked each student to prepare a chart with sketches, poem and/or picture of their chosen significant object, place, thing and person. We further asked them to note what they would do if this unique place or object was threatened in any way. In the third session we engaged with students who had prepared the charts. In Figure 14.1 a student represents the site plan of an old school in his neighbourhood. The trees, playground and building held good memories but were now under threat of demolition.

At the college we interacted with 20 students, commerce and science undergraduates in their second and third year and a mixed-gender group of youth wing members of a community service organisation aged between 19 and 20. The choice of college was, again, deliberate. It is one of the older, well-established city institutions located in the heart of a historic cultural quarter. The first activity was the exploratory interactive session, followed by a group discussion after a week. Finally, we met some students in the field, at their site of choice, in an attempt to encourage them to acquaint outsiders with the significance of the place. Our hope was that, in the future, they might be able to lead visitors around the site and describe its uniqueness. Figure 14.2 shows how a private-school student represents a potter in her neighbourhood.

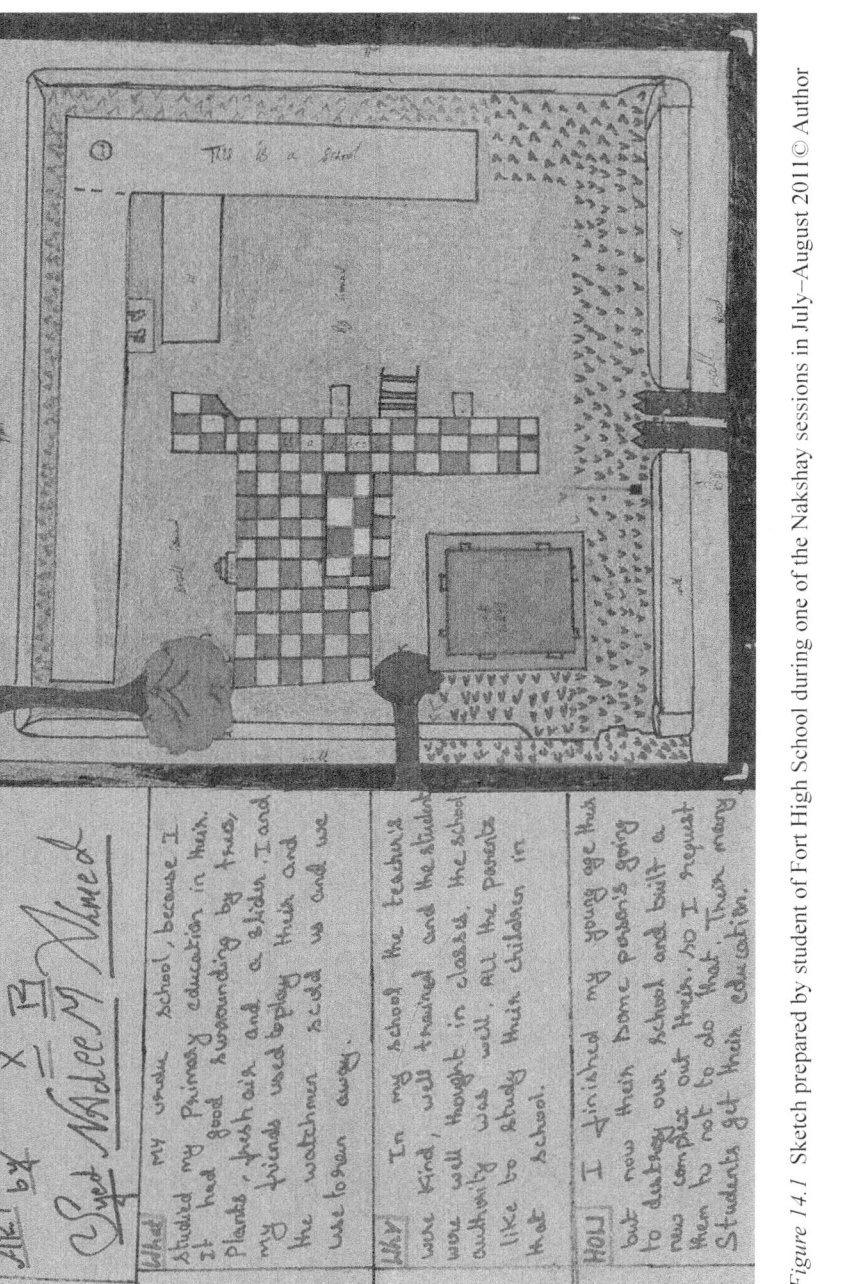

ART by $\frac{X}{=}\frac{\square}{\square}$
Syed NADEEM Ahmed

What My whole school, because I studied my Primary education in this. It had good surrounding by tree, Plants, fresh air and a skidin. I and my friends used to play this and the watchman scold us and we use to run away.

Why In my school the teachers was kind, will trained and the students were well thought in class, the school authority was well. All the parents like to study their children in that school.

How I finished my young age this but now their some person's going to distroy our school and built a new complex out this. No I request them to not to do that. Then many Students get their education.

Figure 14.1 Sketch prepared by student of Fort High School during one of the Nakshay sessions in July–August 2011© Author

Figure 14.2 Sketch prepared by student of Head Start Academy during one of the Nakshay sessions in November 2014. © Chanakya Vyas, reproduced with permission

Neighbourhood Diaries: stories and histories of neighbourhoods

Neighbourhood Diaries is also a personal initiative. It arose out of our desire to focus on local areas. We felt the need as the prevailing view of the city of Bangalore (Bengaluru) is that of a futuristic one, an aspirational place in keeping with its aspirational residents. In the process, the city's history is largely forgotten or assumed to not exist. Thus, our project arose out of a sense of loss or perceived lack of identity among and of the city's historic neighbourhoods. We decided to observe, document and present people's views on local places of socio-cultural significance. Given the dominant view of heritage noted earlier, the underlying focus was to demonstrate that heritage could also be what people value and not just monuments or temples. That is, it could also have social value and not just aesthetic, historical or scientific values.

Here, again, the aim was to instil community pride and foster local identity by creating awareness of cultural heritage. The objectives were to capture and present the essence of rapidly disappearing place history and create strong public awareness of local people's concerns over rapid place transformation.

Partly influenced by the fact that we would be engaging with oral history sources, we ensured that our approach was rigorous. There is, of course, some debate on the veracity of oral history and its use as a valid source of information, but this is not explored in this chapter. Our methods included both archival and field research. We undertook archival research to understand a neighbourhood's documented history and heritage, and field research to identify neighbourhood groups and individuals to engage with. Our next step was to engage different residents in in-depth conversations. This was in order to understand which spaces area residents and users considered culturally significant. Here, again, we used operational questions and asked people about their history of residence in the area, local anecdotes, personal reminisces and so on. We did not directly ask them what they valued. We then walked around the neighbourhood with those individuals we considered had an extensive knowledge of local history and were willing to engage with us. Based on the above methods, we prepared a list of socio-cultural hubs or landmarks we (including local people) considered culturally significant and conceptualised short films around them. Crucially, we limited our choices to places undergoing rapid transformation. We filmed residents in that space, narrating its history, personal anecdotes and their reactions to its rapid transformation. Finally, we presented the films online and in the neighbourhood to encourage a discussion on place significance.

The first neighbourhood we documented was Whitefield. Our aim was to document the area's oral history related to its built tangible heritage (Rajangam 2011). Hence, we presented stories related to significant local places like the church (under threat of demolition due to road-widening), bungalows (under threat of commercialisation) and so on as Whitefield Diaries. The second neighbourhood we documented was Shivajinagar, which we presented as Blackpally Diaries, using the area's colonial-era name. Our work here began in response to the threat of demolition of a single-screen cinema and market in the neighbourhood. However, unlike Whitefield, this project was neither planned nor funded. Second, the area is a commercial hub with a high density of floating population close to the Central Business District (CBD) – so called 'general public'. In contrast, Whitefield is predominantly populated with upper middle class or rich; a desirable city suburb composed of so called 'informed citizens'. However, we found both groups equally difficult to engage with. With Shivajinagar, the issue was more of lingering suspicion over 'what we were getting out of the project'.

Whereas with Whitefield, though we overcame this suspicion fairly quickly, the difficulty was that residents wanted to be helpful: that is, to provide us with the right answers. We faced similar difficulties with *Nakshay* as well.

Further, our approach to both initiatives was also coloured by our own personal bias of viewing *Nakshay* as a research initiative and *Neighbourhood Diaries* as a community outreach exercise. Hence, we assumed we would learn more from the former than the latter. However, as discussion shows in the following section, this expectation did not quite turn out that way.

Discussion

Similarities and differences

Both initiatives engage with communities and attempt to observe, understand and present their views on cultural significance to a wider audience. Here we understand community at its most basic to be a 'set of relationships' – some kind of shared connections or fellowship among a group of people (Williams 1976: 76). The nature of such a connection is perceived to be more organic or natural as opposed to one that is set up formally (Yudice 2005). We more or less stuck to the above definition in *Neighbourhood Diaries*. However, with *Nakshay*, we also chose to interact with communities as defined by demographic. For example, besides groups like residents of Whitefield and so on, we also defined groups as school children, teenagers and so on.

We commenced both projects in 2010–11 on the basis of an initial grant and have since continued to run them as personal efforts without external funding. Further, we chose the internet as a means of dissemination for both. This was in order to make both the project process and results publicly accessible with minimum effort in terms of time and cost.

Considering differences, in *Nakshay* our role is more active. We proactively seek out specific groups that we would like to engage with. For example, following our sessions with the government school, we sought one with private-school children (a different demographic) to understand if a sense of place would vary significantly across the groups. With *Nakshay* we see our role as catalysts towards enabling a discussion on socio-cultural significance of people, places or things. The process is meant to celebrate place and community, and reinforce a sense of shared identity. In contrast, in *Neighbourhood Diaries* our role is more passive; we do not seek out specific communities or groups to engage with. We use the lens of cultural heritage to observe and record how a particular neighbourhood deals with rapid change and potential loss of identity. Thus we also present tensions and conflicts and might, on occasion, facilitate the valorisation of cultural spaces.

Strengths and weaknesses

Thinking about strengths, we can consider positive aspects of the project. Uniformly across the groups and individuals we interacted with, first of all, the fact that someone wanted to record their story or understand their views was regarded as novel. As one of them phrased it, 'this has never happened before'. Second, the project process and results are now all in the public realm and, third, they have benefited from their multidisciplinary nature – our teams have included architects, conservation architects, filmmakers, history students and interested collaborators from other disciplines. Finally, it can be considered that, as a partially funded or non-funded initiative, this is one that runs on passion and personal commitment.

The last aspect can also be viewed as a weakness or negative aspect since it affects the scalability and sustainability of both initiatives. A more significant weakness is our currently unresolved dilemma between objectivity and subjectivity. How objective can we or should we be with *Nakshay*? Our focus is to understand people's views on cultural significance and not impose our understanding on them. Where do we draw the line between initiating a discussion on cultural significance and influencing them with our own views? For example, in the private school, in spite of many sessions, we were unable to proceed beyond contemporary spaces like malls, sporting goods stores and bakery chains. We debated showing the group images of village 'kattes' (informal gathering places), water bodies and vernacular structures, which we knew they encountered en route to school. On reflection, we realised that such contemporary spaces were the cultural frames of reference for these students. They were unaware of a world beyond home (gated community or apartment), school and mall. Thus, rather than consider the sessions a failure on our (or their) part, we learned to accept that not all groups had similar frames of reference. All we could do was make an attempt to expose them to another 'world'.

Impact

Over the years it is apparent that there has been some impact. However, the depth of impact depends on time and the socio-economic-cultural background of the groups or people we engage with. *Neighbourhood Diaries*, surprisingly, had more tangible and immediate results, surprising given our initial bias of considering it an outreach exercise, as noted above.

Short-term or immediate impacts of *Nakshay* included a change in the individual or group's viewpoint over the course of the sessions. For example, a student who mentioned a local mall as a landmark in the first session moved on to describe a historic garden in the final one because it was 'not just a park but a collection of many different things – buildings, flora, fauna...'. Similarly, with *Neighbourhood Diaries* we enabled residents to fight the threat of demolition of a church in one neighbourhood and of a historic market in another.

In partnership with residents, we initiated neighbourhood walks to sites or locations highlighted in both initiatives. Interested locals led the walks, welcomed people to their homes and spoke about their neighbourhood's history. Unexpectedly, this led to awareness creation not just among outsiders but area residents as well. More local people came on board, wanting to know about their neighbourhood and its history – which we did not anticipate.

In the mid-term, the initiatives, particularly *Neighbourhood Diaries*, led to a revival of ties and sense of common local identity. For example, the latter brought together the defunct market association and traders to fight to save their market from demolition. In Whitefield it led to renewed interest in being part of the local Resident Welfare Association.

In the long term, there was awareness creation and promotion of the heritage or historicity of such locations among the general public through media (print and online) coverage. It led to the creation and addition of value to certain significant places or things which may not have been perceived as cultural heritage – for example, an old single-screen cinema that had shut down. Over time it also became clear that the groups we engaged with saw us as a conduit to access information on city-level planning laws, cultural heritage policy and other expertise. They seek our opinion and sometimes advice on different courses of action to pursue when local places are threatened. In effect we were learning from each other, our partnership had become mutually beneficial. An unexpected long-term impact was that heritage became the talking point for different neighbourhood communities. Where there was no common thread, it became a way to bind disparate groups divided along the lines of insider, outsider, migrant, local and so on.

Learning

Methods

As part of our effort to engage with people and groups in a more meaningful way, we adopted different methods – for example, mapping, which was literally getting groups and individuals to mark local cultural hubs on large-scale local area map printouts or online on Google maps. Another method was to have students make charts featuring local hubs. Elsewhere we walked around the neighbourhood with individuals. They pointed out significant places and anecdotes or stories related to these places. We also explored the medium of photographs – residents took pictures of things or places they considered culturally relevant. Finally, in one neighbourhood we trialled the app developed by the Smithsonian Museum project 'Stories from High Street' (with their permission). Residents downloaded the app on their smart phones or tabs and made audio recordings of local place stories. We ensured that we did not conduct formal interviews but always engaged the group or individual in casual, informal conversations.

Of the various methods, the charts and map printouts were most successful. With the others we were limited by lack of access to technology – for example, internet, cameras, smart phones or necessary permissions. Engagement with students was limited to within school premises.

Challenges in engaging people

Being 'local' – that is, a Bangalore resident and knowing the local language – was both a barrier and channel to communication. For example, being considered an 'outsider' – that is, not from their neighbourhood – proved to be an advantage in *Neighbourhood Diaries*, since residents were more open and ready to share local history and personal anecdotes. In some locations it was a barrier: 'You are also from Bangalore, right, you must know all this... what more can we tell you?'. With *Nakshay*, as our initial engagement with the group was through more formal channels, introductions mattered, namely who introduced us to the group and how.

Genuine effort to engage people, including the fact that we truly wanted to work with them through the whole process, met with success. For example, with the films, we ran the edited version past local residents who contributed to it, sought their clearance and then completed it. On occasion, we have changed content, based on participants' discomfort with how their views were expressed on screen. Once people realised that we were not 'getting anything out of it' they opened up. However, the time and effort involved in breaking the ice varied. For example, the market traders were more suspicious of our motives in engaging with them and our wish to present their story. Unlike them, the residents of the Anglo-Indian community accepted us a lot more quickly, though they were also suspicious initially.

In general, we found adults the most difficult group to engage, unless they were already so inclined or shared our interest. For example, working with a group of affluent adults we realised that while they were happy to spare an afternoon or couple of afternoons to map places with local cultural significance, they saw the activity more as helping us with a research project rather than something they would continue to think about. However, the level of awareness among teenagers did surprise us. They showed a concern for the environment, loss of green space and trees and increased pollution.

What do people value?

Very briefly, what *did* people value? No common thing or place as such, it depended on the group. However, we did notice a tendency for things or people or places with personal associations to be valued over other culturally significant properties. In general, natural heritage was valued over built heritage.

School-student responses ranged from temple processions to their own house. College-student responses to culturally significant places ranged from large living temple complexes to the sprawling tree-covered campus of a research institute. Responses tended to reflect the group demographic – for example, with students of the private school mentioned earlier, we were unable to progress the discussion beyond their cultural frames of reference. School-student responses referenced places geographically closer to home, unlike college students, while the latter's responses also included 'local hangouts'.

Finally, while cultural heritage took on the role of a binding instrument, it is not possible to say whether it was just a convenient rallying point at that moment or whether it truly contributed to a common sense of well-being. What became quite apparent is that cultural heritage can only be included as part of a larger agenda, whether education or culture or sustainability. It was never seen as a central concern by itself.

Conclusion

In conclusion, both initiatives have had an impact and they have tended to be positive in their outcomes and effects. The tendency has been to celebrate place and community rather than capture conflicts and tensions. In some instances they have managed to foster local identity and pride in their surroundings. But we do not know how long such effects lasted.

However, the initiatives are significant because they are genuine attempts to engage communities in a dialogue on heritage and its conservation. While, in popular literature, heritage and conservation continue to be seen as something to do with monuments and their scientific management, these initiatives become an opportunity to foreground people's views on cultural places. In one sense they are novel and useful attempts to bridge gaps between official policy and local people's viewpoints. In a country like India, with vast unprotected heritage, such initiatives are a way to document views that might otherwise go unheard or remain unrepresented.

One future direction for both projects would be to continue to test methods that enable maximum participation. A larger vision would be an attempt to measure their impact and think through the place of socially driven conservation within the larger field of conservation studies. Second, I hope the initiatives will enable a discussion on the nature and role of cultural heritage in a particular context. I consider this would be a useful contribution to the field of heritage studies in India.

Acknowledgements

Thank you *Nakshay* and *Neighbourhood Diaries* teams – architects Pankaj Modi, Sonalika Dugar and Aparna Shastri; filmmaker Clemence Barret; and the communities who are at the heart of both initiatives for their continued support. *Nakshay* owes much in its approach to the UK national charity Common Ground, which champions democratic involvement in localities. The concept of creating awareness about local places was influenced by an interview with the founders of Common Ground. Thanks are also due to Susanna Batel, Geography Department, University of Exeter. Her interview with me as part of their research on the Parish Maps project of Common Ground helped me to think through both initiatives clearly in terms of strengths, weakness and impact. This conversation helped me develop the outline for the discussion section of the current chapter.

Notes

1 I interpret 'cultural heritage' as a thing, place, custom or person that a particular group or community considers socio-culturally significant, and 'conservation' as a process

(basing it on the *Burra Charter* definition). While the prevalent view among authority is that of a technical process, views within the conservation discipline vary.

2 www.nakshay.saythu.com and www.neighbourhooddiaries.wordpress.com

References

Ashley, K.S., Osmani, M., Emmitt, S., Mallinson, M. and Mallinson, H. 2015. 'Assessing stakeholders' perspectives towards the conservation of the built heritage of Suakin, Sudan'. *International Journal of Heritage Studies* 21(7), 675–97

Bunten, A.C. 2008. 'Sharing culture or selling out? Developing the commodified persona in the heritage industry'. *American Ethnologist* 35(3), 380–95

Collins, J. 2008. '"BUT WHAT IF I SHOULD NEED TO DEFECATE IN YOUR NEIGHBORHOOD, MADAME?": Empire, Redemption, and the "Tradition of the Oppressed" in a Brazilian World Heritage Site'. *Cultural Anthropology* 29(2), 279–328

Fritz, J.M. and Michell, G. 2012. 'Living heritage at risk'. *Archaeology* 65(6), 55–62

Gentry, K. 2014. '"The pathos of conservation": Raphael Samuel and the politics of heritage'. *International Journal of Heritage Studies*, DOI: 10.1080/13527258.2014.953192

Harvey, D.C. 2001. 'Heritage pasts and heritage presents: temporality, meaning and the scope of heritage studies'. *International Journal of Heritage Studies* 7(4), 319–38

Rajangam, K. 2011. 'Whitefield: an important but forgotten chapter of India's colonial heritage'. *South Asian Studies* 27(1), 89–110

Rajangam, K. and Prasad, A. n.d. *Neighbourhood Dairies* [online]. Available at www.neighbourhooddiaries.wordpress.com [accessed 27 August 2015]

Saythu 2011. *Nakshay* [online]. Available at http://www.nakshay.saythu.com/ [accessed 27 August 2015]

Thakur, N. 2012. Keynote address, Pracheen Tatva, National Conference on Architectural Heritage Conservation and Management, Department of Architecture, MSR Institute of Technology, Bangalore. 18–19 April

Williams, R. 1976. *Keywords: A Vocabulary of Culture and Society*. London: Fontana/Croom Helm

Yudice, G. 2005. 'Community'. In T. Bennett, L. Grossberg and M. Morris (eds), *New Keywords: A Revised Vocabulary of Culture and Society*. Malden, MA, and Oxford: Blackwell

15 Community involvement matters in conserving World Heritage Sites

Remote cases of Japan

Aya Miyazaki

Introduction

The UNESCO *Convention Concerning the Protection of the World Cultural and Natural Heritage* (hereafter, the Convention) established in 1972 has greatly changed international perception towards cultural heritage conservation. World Heritage inscription has become an international symbol that aims to protect significant cultural and natural heritages from losses and preventable damage. And yet, its mechanism is not understood well by those outside the conservation field. The accepted understanding of World Heritage Site (WHS) is that once a site is placed on the World Heritage List, the work is done. However, this is not a pageant of cultural and natural sites, but an international mechanism for heritage preservation through a dual process of inscription – that is, formal designation and conservation management. The state party's government, as a signatory to the Convention, is responsible for protecting the site under its national statutory conservation system after inscription.

Diverse types of cultural heritage from around the world are thus conserved in a single scheme, using an expert-led concept of Outstanding Universal Value (OUV), as defined in the *Operational Guidelines for the Implementation of the World Heritage Convention* (UNESCO 2013). Each state party is given the flexibility to manage the WHS according to its own methods as no specific conservation regulation is written into the international documents. Thus, the conservation approach and the level of site management will vary according to each nation's legislation and policies.

The communities of WHS are another matter. They attach specific emotions and values to a site regardless of its national or international designation. Communities as actors, however, have largely been absent from the international documents and national legislations.[1] What is more, the structural complexity of the WHS system, with multiple layers of cultural heritage values – from local, national to international (OUV) levels – has often been overlooked.

This chapter focuses on the different layers of values that comprise the concept of a WHS and how the 'community' (as defined below p. 235), is involved in the management of a site, given its direct role in this after the World Heritage inscription. By focusing on Japanese case studies that show collaboration with local communities at an early stage of conservation, this chapter will suggest that such early community engagement is important for an effective protection of the WHS in the long term. This research (part of a Conservation Studies MA at the University of York) was conducted in conjunction with communities in Japan, interviewing local representatives for their views and perceptions about conservation of cultural heritage.

Conservation mechanism and values of World Heritage Sites

The World Heritage conservation mechanism

The first phase of the dual conservation mechanism for WHS is the process of inscription – making and evaluating nomination dossiers submitted by state parties, prepared after a meticulous process of site research, documentation and development of conservation/management plans within the national statutory system. After careful evaluation of the dossier by UNESCO's advisory bodies, the result is reported to the World Heritage Centre and the decision to inscribe, postpone, or defer the nomination is finalised by the World Heritage Committee (representatives from 21 of the state parties to the Convention elected by the General Assembly) at its annual meeting.

It then shifts to the second phase of the mechanism: conservation — where state parties and related actors take responsibility for long-term site management. This is based on the submitted conservation management plans within the framework of their national legislations, following advice from the WH Committee's Advisory Bodies (IUCN, ICOMOS and ICCROM). In ratifying the Convention, the government undertakes to protect all national heritage, regardless of its WHS or a national designation (art. 4, 5), and assist other state parties in their conservation efforts (art. 6, 7).

Nevertheless, however carefully a WHS is managed, conservation issues may still be difficult to resolve, as indicated by the number of 'State of Conservation' reports to the Committee each year (UNESCO 2015). When threats convert into imminent dangers of permanent damage to the site, the process shifts to a second stage of conservation — World Heritage in Danger enlistment. There is a continuous cycle of surveillance by the international community, intervention by the Committee, active monitoring and assistance of the experts, and international attention that leads to a global support; it functions as a key to encourage active governmental conservation and protection of the OUV. Ultimately, however, the only sanction is removal of a threatened site from the World Heritage List. The sovereignty of individual state parties permits flexible management of the WHS based on the national statutory system of each government. Although such flexibility has gathered international support for the OUV concept, widened doors for the cultural heritage protection and helped develop national conservation systems around the globe, it has resulted in diverse site management approaches and, in some cases, divergent standards of conservation.

Values

Thus, such freedom of state parties to decide on the conservation practice has created a rather complex set of values to be applied in site protection and management. At least three levels of value sets are currently used to assess a WHS: international, national and local.

The conservation mechanism can be said to consist of both political and professional aspects. An epistemic community provides expertise and applies a certain set of values to cultural heritage conservation. The cultural heritage, however, is also selected based on political biases and governmental decisions.

The Convention, linked to the highest level of the three-tiered value set (Figure 15.1), recognises 'heritage' based on the concept of the OUV, defined by expert advisers as 'cultural and/or natural significance which is so exceptional as to transcend national boundaries and to be of common importance for present and future generations of all humanity' (UNESCO 2013, art. 49). The OUV must meet at least one of the ten criteria as well as the conditions of authenticity and integrity defined in the Operational Guidelines, assessed according to professional values. Under the Nara Document principles, now attached as an Annex to the *Operational Guidelines* (UNESCO 2013: II.E.79), a holistic definition of authenticity aims to ensure a full range of cultural heritage diversity being encompassed in conservation planning. As Sian Jones's (2009, 2010) research in the UK has demonstrated, authenticity can be defined through relationships and dialogues between the person and the thing or place and is not limited to materiality or the cultural significance attached to the object.

Such is the case at the World Heritage Committee, where sometimes experts' evaluations are altered after political negotiations and discussions among the Committee member countries. With much lobbying and political pressure of the state parties aiming to inscribe a site, it is not only the cultural significance but also political pressure on international governmental decision-making process that has created the World Heritage List.

Figure 15.1 Three-tiered value pyramid of the WHS. Diagram © Author

Similarly, national heritage values show binary aspects of heritage. Nationally designated sites are evaluated and maintained by heritage experts but WHS are selected according to political interest or national criteria, differently in each country. As each state has the freedom to set a unique selection process and criteria, its 'best' heritage, based on political, cultural, economic, or religious motives, is chosen, which will be reflected in the tentative list prepared for the inscription of the WHS.

The third level — local value — contains ideas of authority and community that are essential in grasping the different perceptions between the local constituencies in cultural heritage. Aside from the political/professional binary, value-setting process at the local authority level, the closest people to the site — the community – has a different perception towards the site. As expressed in the Nara Document (ICOMOS 1994), everyone has a personal connection and an attachment to a history and heritage, on which there is not necessarily mutual agreement by others – a distinctive characteristic of 'insiders' in the categorisation of heritage stakeholders distinguished by Howard (2003). There is a contrast between insiders and outsiders (those external to the paradigm of insider value but concerned with access and interpretation of cultural heritage as tourist, educational visitor, pilgrim or connoisseur) in communities. The core 'insider' group engaged in conservation issues may be those that place conservation at the centre of their lifestyle for occupation or may be those active in conservation planning processes for personal interest/attachment in order to 'control space', as suggested by Pendlebury in his analysis of people's engagement with conservation issues (2009: 142). Such heterogeneous perceptions within a community make interpretation of communal values complex and contested.

Consequently, a World Heritage Site presents a mélange of conservation management, engaging with three different sets of complex values: those of the international, the national and the local community. For the purposes of this chapter, community is defined as a group of individuals with personal attachment to, or affected by, the WHS. The community may be residents and employees in or near the designated WHS, or those with a special interest or concern about the particular site, such as NGOs/charities, amenity societies, epistemic or other community with a special knowledge or experience of the place.

Conservation institutions and legislation in Japan

Japan ratified the Convention in 1992, and WHSs in Japan are conserved under the *Law for the Protection of Cultural Properties* (hereafter, the Protection Law, ACA 1950) enacted in 1950 as the binding statutory system for all tangible and intangible cultural properties. Supplemented by the cabinet orders and regulations of local government for specific actions, the Protection Law replaced all previous laws dating back to 1897 (Nakamura 2007: 1).

The objective of the Protection Law (ACA 1950) is to 'preserve and utilise cultural properties, so that the culture of the Japanese people may be furthered and a contribution be made to the evolution of world culture' (art. 1). The

Protection Law states the duties of the national and local governments (art. 3), including their responsibilities to teach the importance of cultural properties to the public and the property owners, as well as to respect the ownership of the property (art. 4). As Larsen (1992: 4) has observed, two significant characteristics of Japan's Protection Law are its emphasis on the public's right to access cultural properties and its philosophy of including a variety of cultural and natural properties in a single statutory system.

Cultural properties are distinguished into six categories according to the criteria of materiality/immateriality, single/groups of properties, artistic/vernacular usage, artificial/natural context and organic/inorganic objects: 1) tangible cultural properties (including buildings), 2) intangible cultural properties (art and skills in traditional performing/applied arts), 3) folk cultural properties (tangible and intangible), 4) monuments (from sites to animals, plants and geological formations), 5) cultural landscapes (in association with the modes of life) and 6) groups of traditional buildings (such as port/castle towns, farming/fishing villages). While historic, artistic, academic and age values underlie the Protection Law (ACA 1950), age value is especially significant for the historic buildings. Most of the surviving architectures of the Asuka, Nara and Heian periods (538–1185 AD) are enlisted either as Important Cultural Properties or National Treasures (Nakamura 2007).

The degree of national significance of a cultural heritage site affects the levels of governmental involvement, legal responsibility of the property owners, and penalties for improper heritage management (ACA 2014). The Minister of Education, Culture, Sports and Science *designates* nationally important tangible, intangible and folk cultural properties as Important Properties and universally significant sites as National Treasures. Once designated, the property owners could be penalised for undeclared changes made on the property, but will also be able to receive subsidies covering at least a half of the conservation repair expenses. Undesignated tangible properties will be *registered* for management, preservation and public access by local or national government; cultural landscapes and groups of traditional buildings will be *decided by selection* under the local governmental regulations with the consent of the communities and residents (Nakamura 1999: 8–10).

Through four amendments (in 1954, 1975, 1996 and 2004) the Protection Law has been adapted to the changing environment of historic conservation caused by social and economic factors in Japan. Major structural change was the result of the economic boost in the 1950s–60s (ibid.: 22–4) which instigated civil movements in rural towns to preserve historic buildings from destruction, raising the significance of communities in conservation. As people moved out of the provinces for jobs in the cities, depopulation became an issue in the remote areas. In contrast, rapid urbanisation of the cities altered the urban landscape on a large scale through demolition of traditional buildings and development of modern infrastructure and architecture.

Such issues, in both urban and remote regions of Japan, led to the addition of the Groups of Traditional Buildings in the Protection Law in 1975. A bottom-up approach is taken for this cultural property to protect the façade of the historic architecture and the traditional landscape. Community agreement must be reflected in the municipal regulations to control the development and protection of traditional buildings, while the state government chooses nationally significant districts to provide financial support to protect the cityscape.

Although its significance is acknowledged, community is yet to be considered as a key player in the current Japanese cultural heritage conservation system. Individual owners – whether private or institutional – are primarily responsible for fundamental administration, restoration and the public display of the cultural property, but significant administrative decisions and regulation are 'top-down' through the governmental and professional bodies of the Ministry of Education, Culture, Sports and Science (MEXT) and its auxiliary department, the Agency for Cultural Affairs (ACA). Orders on designated cultural properties are made by either the Minister of MEXT or the Commissioner of ACA and are reviewed by the Council for Cultural Affairs, the specialists of the cultural properties with a significant impact on making decisions at the national level (Nakamura 2007: 31). Similarly, the Council for Preservation of Cultural Properties (Bunkazai–Hogo–Shingikai) is placed under the Board of Education, part of the bureau that functions distinctively from the rest of the administrative office to consider conservation and management issues at prefectural and municipal levels, for the administrative role (ACA 1950, art. 190).

World Heritage Sites in remote areas

Turning now to the remote cultural heritage, we can consider the historic villages of Shirakawago and Gokayama and Iwami Ginzan silver mine and its cultural landscape to analyse community involvement in conservation of cultural heritage. Remote WHS are defined here as cultural heritage located outside an urban centre in an area with a comparatively smaller population. Residential areas within a rural district are included but the number and range of communities are considered smaller and more restricted than in the urban heritage.

Depopulation from the rural areas into cities, for the job opportunities created by economic growth in the 1960s, initiated strong community movements to protect their ancestral heritage and traditional landscape for both of these case studies of remote cultural heritage sites. Municipal bodies paid attention to such community movements and designated these sites nationally under the Protection Law, which were later inscribed as WHSs.

Historic villages of Shirakawago and Gokayama

Shirakawago and Gokayama were villages inscribed in the World Heritage List in 1995 under the criteria of (iv) and (v) (UNESCO n.d. a). The combined core zone area of three villages is 68 ha and the buffer zone is 58,873 ha in total, spreading across the two prefectures of Gifu and Toyama. Known for their long period of heavy snow in the secluded deep mountains, the villages share a unique building structure of a steep thatch roof to slide off snow, called the *Gasshozukuri*. The high ceilings of *Gasshozukuri* enabled silkworm breeding that became the key industry of the region. This chapter examines Shirakawago, one of the predecessors of community-driven conservation of cultural properties in Japan.

Shirakawago is a village of 1,733 people (as of 2010), with 601 households divided into 14 districts. The only district with surviving *Gasshozukuri*, Ogimachi district was nationally designated as one of Japan's first Important Preservation Districts for Groups of Traditional Buildings in 1976 by the ACA. Regulations of the Important Preservation Districts, and a master plan without statutory power, serve as the guideline for development and conservation of the history of the village (Matsumoto 2013).

Ogimachi divides its 150 households into seven Kumis (groups) to re-thatch roofs every three to four decades in a traditional manner of Yui system; all villagers in a Yui group gather to fully re-thatch a roof in their neighbourhood in a single day. *Gasshozukuri* rapidly disappeared during the economic boom, when villagers migrated from the neighbouring villages to the cities, leaving houses without maintenance. Later, dam construction for hydro-electricity engulfed villages nearby to sustain urban energy demands.

Villagers of Ogimachi established their Protection Society and Residential Charter in 1971 under a strong leadership of the then chief of the village to confront the imminent threats. Three principles of 'Not selling', 'Not renting' and 'Not destroying the *Gasshozukuri*' underlie the Residential Charter to protect their buildings' use, as well as their nineteenth-century traditional architectural style (Protection Society 2010: 15).

The Protection Society, the key community body in charge of conservation, meets once a month to discuss applications for changes and conservation of the *Gasshozukuri* before decisions are finalised by the Board of Education. Each Kumi sends a two-year-term representative to the meeting, where the villagers learn the importance of protecting their traditional architectures and the philosophy behind it. The Society has seen increasing representation of younger

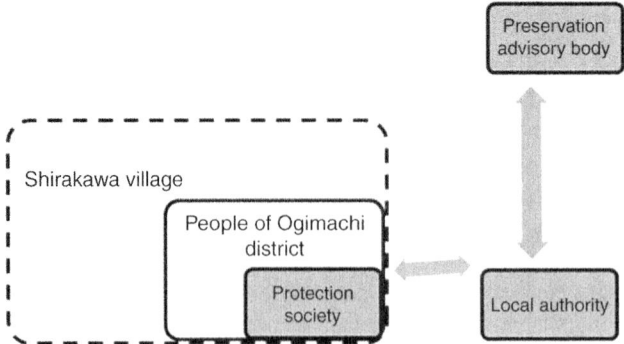

Figure 15.2 Communities affecting decisions about management and conservation at Shirakawa village. Diagram © Author

members at its meetings in recent years; they are being educated by the elders about their unique history. Mr Matsumoto, representing the Board of Education, mentioned in interview that the central actors of conservation and decision-making are the community members themselves. By balancing out the autonomy of the Protection Society and involvement of the local authority, the community members are capable of convincing their fellow residents to follow the three principles and protect their ancestral heritage (Matsumoto 2013).

Limited to a supporting role, the local authority's involvement is significant to enable community-driven conservation. When there are complex issues requiring professional perspective, the matters will be taken into account by the Preservation Advisory Body, formed of cultural heritage professionals. For minor repairs that cannot be subsidised by the government, a three-hundred yen fund established by the Shirakawago village and Gifu prefecture is appropriated to financially aid the property owners. In addition, a four-body authority meeting with Gokayama village and Toyama prefecture was established in 2010 to coordinate a holistic approach to the World Heritage management.

It was not until 2008 that the official guidelines for mass tourism and transportation traffic were introduced to cover areas that are not controlled by the regulations of the Groups of the Traditional Buildings. Community meetings to propose a management plan started with expert advice in 2007. Thirty members met 16 times and formed a vision of Ogimachi, three columns of conservation management and seven guidelines (Shirakawamura 2010). The final proposal was presented at the annual end-of-year Ogimachi meeting, where all 150 households send a family representative to discuss important matters. After the community consent, the proposal was submitted to the Board of Education to be utilised as the foundation for the village master plan. The Steering Committee, consisting of the Board, ACA, experts and two community members — the Head of the Ogimachi district and the President of the Protection Society — discussed the duties and responsibilities of the authority and established a master plan in 2010, based on the village proposal. It now functions as the foundation for future development and conservation of the town.

A 'Tourism Action Plan' was also enacted in 2013 to deal with increased external pressures caused by a sudden increase of visitors. After the designation as Important Traditional Buildings, 600,000 tourists visited in 1976; the number greatly increased to 1.3 million in 1995, after the inscription of the WHS, and hit a ceiling of 1.86 million visitors annually in 2008. A major issue was the conversion of small rice paddies in front of the *Gasshozukuri* that were turned into parking lots to earn tourist money, which started to affect the landscape of the village. After long negotiations among the Protection Society and the car parking space owners, vehicles were restricted entry into the core zone of the WHS from the end of 2013 and the parking spaces are limited to the village-run car parking outside the main *Gasshozukuri* area.

The strongest sense of community, however, is restricted to that of Ogimachi, the core area. The other six districts within the Shirakawago village that compose the buffer zone feel excluded from the WHS. As a non-Ogimachi guesthouse owner stated in an interview, there is a sense of insider/outsider within the Shirakawa village, between Ogimachi and the others, because of the lack of information being shared among Ogimachi residents and residents of other districts (Minoriya 2013). Even within the Ogimachi district, there is a growing economic gap between those who earn tourist money and those who are not affected by it, creating a sense of insider/outsider in the core zone as well.

Iwami Ginzan silver mine and its cultural landscape

Iwami Ginzan was inscribed in the World Heritage List in 2007 under the criteria of (ii), (iii) and (v) (UNESCO n.d. b). It is a holistic, cultural landscape of an archaeological and historic silver mining site in its natural environment, which developed and declined with the silver mining industry. The core and buffer zones are located within Oda city, which is the main community of the Iwami Ginzan, with a total population of 2,000. There are multiple elements to the WHS, from below-ground archaeological remains to routes that transported silver from the mountains to the port. Shimane prefecture and Oda city function as the preservation management organisation that legally control development and protection, on behalf of multiple owners according to the Protection Law (ACA 1950, art. 113). The two authorities form the Conservation Management Council to prevent inappropriate development and mal-conservation by serving as a management representative of the area at the administrative level.

The WHS consists of four districts: Omori, the main component of the site where the silver mine is located, and three port towns of Yunotsu, Tomogaura and Okidomari. The biggest populations are gathered in Omori and Yunotsu districts, with 400 residents each, divided respectively into ten and nine self-governing bodies, while only ten residents live in the small port towns of Tomogaura and Okidomari. As Omori is the only area within the WHS where the community initiated a conservation movement before the national designation, the local authority of Oda city considers Omori district as the leading case study for the other three districts to pursue (Oda City Board of Education 2013).

After the silver mine was closed in 1923, meticulous historical research by local historians and community campaigners started by recording and preserving the landscape and history. It led to the creation of Omori Town Cultural Heritage Preservation Society (hereafter, the Preservation Society) in 1957, where all residents of Omori may be a member; it meets at least once a year to discuss conservation matters. The silver mine was designated as a prefectural historic site in 1967 and a National Historic site in 1969; the residential area was designated as Groups of Traditional Buildings in 1987 (Oda City 2012a). The strong community involvement encouraged the Shimane prefecture to nominate the place as a WHS in 1995 (Yasue 2011: 39). The Neighbourhood Council of Omori is another community of residents in charge of managing and deciding matters of importance regarding development, tourism and conservation of the Omori district. Omori's quick reaction to preserve the silver mining site formed an active Preservation Society and vigilant Neighbourhood Council that strengthened conservation action. It enabled the city to cope with the growing number of visitors, increasing from 300,000 after the national designation to 813,200 in 2008 after the World Heritage inscription (Sataki 2009: 143).

A unique movement of the Iwami Ginzan 'Cooperation Council' (trans. author) started in 2005 before the World Heritage inscription, to establish an action plan for management of the site (Figure 15.3). It is a trilateral panel of representatives from business entities, active community groups of Oda City and the local authority as a secretariat of the Council. The Cooperation Council established the Residential Charter, rules for business, a park-and-ride system and infrastructure/construction to sustain the economic benefit of the city. A group of 200 individuals interested in the World Heritage inscription gathered and discussed how the site should be conserved and managed, regardless of the international designation. They met 12 times as a whole and a total of 61 times in five subgroups to discuss conservation and management of Iwami Ginzan in a single year (Oda City Board of Education 2013).

The Cooperation Council is now in charge of a 300 million yen fund for the site management. Half is donated by the general public, Shimane prefecture and Oda city donated 75 million yen each to complement the 300 million yen fund. By pooling money to use for conservation of the silver mine, the local authority eschews utilising a regular annual budget for the site management (Oda City 2012b).

Unlike Omori, the other three districts of Yunotsu, Tomogaura and Okidomari were included later in the nomination scheme to create an integrated approach to the region's silver mining history, after this was suggested by expert advisers of ICOMOS. Thus, Boards of Education of Shimane prefecture and Oda City, instead of the communities, have been the central actors of conservation for these areas, nominating port towns to be designated as Groups of Traditional Buildings in 2004 and 2005 (Oda City Board of Education 2013). Lack of community leadership in

Figure 15.3 Collaboration Council scheme of Iwami Ginzan. Diagram © Author

the process resulted in the indifference of these later-added communities towards the site when compared to Omori's residents. Although the four districts compose a WHS, the latter three districts have weaker economic gains from tourism due to their distance and lack of efficient public transportation system from the Omori district. Such indifference is clear in the response of a 90-year-old lady whom I interviewed at the Yunotsu hot spring. When I asked her of how the World Heritage designation impacted Yunotsu, her reply was 'No, Yunotsu is not a World Heritage site. Omori is' (Yunotsu community 2013).

Conclusion

Community involvement in conservation prior to national and international designation as a WHS has served as an engine to manage sites effectively in remote areas and to create a holistic conservation approach. The case studies show that the homogeneity, smallness and closeness of the core group have formed explicit and straightforward action plans that can be shared with the local authorities. Motivated by a strong anxiety about the loss of their ancestral past due to depopulation, the communities took charge of protecting their cultural heritage sites and were active in creating their own set of conservation rules as well as places where their voices could be reflected in the municipal management and conservation plans later on.

The local authorities have successfully utilised such community movements and reflected their opinions in the action plans. This resulted in an active involvement of communities in the conservation and management of the remote WHS, creating a context where the communities become significant actors in the cultural heritage management under the current system of national protection. As the municipal governments consider tourism the sole economic means in the region, there is an incentive for them to actively reflect the local voices in actual practice to create a robust economic foundation. The two distinct objectives of local communities and governments formed a strong foundation and sustainable conservation scheme for the two WHSs.

However, the contrast between 'insider' and 'outsider' became apparent even among the WHS communities. Compared with the core groups who initiated the conservation movement, the community members who were later included in the conservation scheme by the state/municipal government do not feel as much a part of the WHS community.

Nonetheless, the World Heritage designation has changed the ways communities look at their places in general. While the national designation controls the ways cultural properties are preserved and managed, international designation has changed the mindset of the community members and raised the community's attention to notice more particular details and issues; it is now the residents that report conservation problems in the neighbourhood to the local authorities. The number of cases in this chapter is limited for significant conclusions, but these two examples illustrate the complex structure of values and actors involved in the conservation of a WHS. Although community involvement is not clearly written into the international documents, their role is essential to achieve a community-centred, practical management of the World Cultural Heritage sites in these remote areas of Japan.

Note

1 As Jokilehto notes in Chapter 2, there has been a sustained trend in international recognition of the community's role in the processes of safeguarding and actively managing heritage resources since the early 2000s. Mechanisms by which this may be achieved are not explicit in the Convention, however.

References

ACA (Agency for Cultural Affairs), Government of Japan 1950. *Law for the Protection of Cultural Properties* [Bunkazai–Hogo-Hou]. Horitsu 214. Adopted 30 May 1950 [online]. Available at http://law.e-gov.go.jp/cgi-bin/idxselect.cgi?IDX_OPT=4&H_NAME=&H_NAME_YOMI=%82%a0&H_NO_GENGO=H&H_NO_YEAR=&H_NO_TYPE=2&H_NO_NO=&H_FILE_NAME=S25HO214&H_RYAKU=1&H_CTG=27&H_YOMI_GUN=1&H_CTG_GUN=1 [accessed 19 April 2015]

ACA, Government of Japan 2014. *Policy of Cultural Affairs in Japan – Fiscal 2014.* Tokyo: ACA

Howard, P. 2003. *Heritage: Management, Interpretation, Identity*. London: Continuum

ICOMOS 1994. *The Nara Document on Authenticity*. Drafted at the Nara Conference on Authenticity in Relation to the World Heritage Convention 1–6 November. Nara: UNESCO [online]. Available at http://www.icomos.org/charters/nara-e.pdf [accessed 19 April 2015]

Jones, S. 2009. 'Experiencing authenticity at heritage sites: some implications for heritage management and conservation'. *Conservation and Management of Archaeological Sites* 11(2), 133–47

Jones, S. 2010. 'Negotiating authentic objects and authentic selves: beyond the deconstruction of authenticity'. *Journal of Material Culture*, 15 181–203

Larsen, K. 1992. 'Authenticity and reconstruction: architectural preservation in Japan'. Unpublished manuscript, Kings Manor Library, University of York

Matsumoto, K. 2013. Personal Interview. Semi-structured interview for MA dissertation. At local authority office in Shirakawa village, 1 August

Minoriya 2013. Personal Interview. Semi-structured interview for MA dissertation. At the guest house in Shirakawa village, 31 July

Nakamura, K. 1999. *Bunkazai–Hogo-Seido–Gaisetsu* (Overview of the System for the Protection of Cultural Properties (trans. by the author)). Tokyo: Gyosei

Nakamura, K. 2007. *Wakariyasui Bunkazai-Hogo-Seido no Kaisetsu* (Easy to Understand Guide to the Law for the Protection of Cultural Properties (trans. by the author)). Tokyo: Gyosei

Oda City 2012a. 'Iwami Ginzan silver mine and its cultural landscape'. Document provided by the local authority of Oda City Board of Education, July 2013

Oda City 2012b. 'History of Iwami Ginzan Foundation, 1. Oda City: Oda City'. Document provided by the local authority of Oda City Board of Education, July 2013

Oda City Board of Education 2013. Personal Interview. Semi-structured, group interview for MA dissertation. At Oda City Local Authority, with Haruo Ooguni, Yasukuni Hayashi and Kenichi Nakata, 29 July

Pendlebury, J. 2009. *Conservation in the Age of Consensus*. London: Routledge

Protection Society (Society of the Natural Environment Protection in Ogimachi District at Shirakawa-go) 2010. 'Ogimachi district at Shirakawa-go: 40 years of history – appreciating, learning from ancestors and connecting to the future generation' (trans. by the author). Shirakawamura: Maki–Insatsu. Document provided by the local authority of Shirakawa-village, August 2013

Sataki, Y. 2009. *'Sekai Isan' no Shinjitsu – Kajou na kitai, Ooinaru Gokai* (Reality of 'World Heritage Sites' – Excessive Expectation and Great Misunderstanding (trans. by the author)). Tokyo: Shodensha

Shirakawamura 2010. 'Shirakawa village World Heritage Site master plan: A plan to inherit the heritage and accomplish a rich life' (translated by the author). Document provided by the local authority of Shirakawa village, August 2013

UNESCO 1972. *Convention Concerning the Protection of the World Cultural and Natural Heritage*. Adopted by the General Conference at its 17th session, entered into force 16 November 1972. Paris: UNESCO

UNESCO n.d. a. *Historic Villages of Shirakawa-go and Gokayama* [online]. UNESCO. Available at http://whc.unesco.org/en/list/734 [accessed 19 April 2015]

UNESCO n.d. b. *Iwami Ginzan Silver Mine and its Cultural Landscape* [online]. Available at http://whc.unesco.org/en/list/1246 [accessed 19 April 2015]

UNESCO 2013. *Operational Guidelines for the Implementation of the World Heritage Convention*, WHC.13/01. Paris: World Heritage Centre

UNESCO 2015. *State of Conservation Information System (SOC)* [online]. UNESCO. Available at: http://whc.unesco.org/en/soc/ [accessed 19 April 2015]

Yasue, N. (ed.) 2011. *Sekai Isan-gaku e no Shoutai* (Guide to the Study of World Heritage Sites (translated by the author)). Tokyo: Horitsu-Bunka-sha

Yunotsu community 2013. Personal Interview. Spontaneous interview for MA dissertation. At the Yunotsu hot spring in Yunotsu port town, 28 July

16 Researching with the public, conserving with the community
The Martos project workshop, Spain

Laura-Melpomeni Tapini and Lucía Gómez-Robles

Introduction

For decades, heritage conservation has been exclusively dealt with by specialists. However, heritage is becoming more and more a subject of general interest. Involving the public in different stages of heritage management is gaining ground and its contribution can actually become the key for creating sustainable conservation models.

In the Martos project, a restoration educational workshop on a sixteenth-century monumental fountain, we involved the local community as a key strategy in three interconnected stages, whereby continuous maintenance throughout the future can be ensured. The present chapter summarises questions, methodology and strategies developed during the preparation and implementation of an international workshop, organised in 2012 by Diadrasis in Andalusia, Spain.

Diadrasis project workshops and new challenges

Diadrasis is a Greek non-profit organisation aiming at the conservation and protection of cultural heritage through the interaction of specialists with different professional and national backgrounds. Diadrasis's activities focus on four main areas: non-formal education, research, public awareness activities and publications. In this framework we have developed 'project workshops' – that is, small-scale educational restoration projects, with the main aim of engaging participants in comprehending all the different tasks of a complete conservation project (Tapini and Gómez-Robles 2013).

Having successfully completed the pilot project workshop of a small chapel in Greece (Diadrasis 2011) and while designing the second one, planned in Spain, we found ourselves facing two new challenges. First, the severe economic recession, a world-wide problem, with great impact on heritage conservation; and, second, the question of 'minor' heritage.

There is no doubt about the impact of the economic crisis on prioritising of state and private funding (Inkei 2010). It is often debated that – in periods of severe human crises, like the current one, provoked by the rise in the rate of unemployment – investing money in culture and heritage is a luxury that society cannot afford. Nevertheless, if we allow the creation of a temporary 'black hole' in cases of conservation and management of our heritage until the financial crisis is over, we run the risk of losing parts of our built heritage and so a part of our history forever (Hooper 2010).

Directly related to the financial crisis and to the richness of built heritage in the countries involved in the project, but not only applicable to them, is the case of 'minor' or local heritage. As the time and subject span of what is considered heritage continues expanding (Muñoz Viñas 2005), the difficulty of managing and maintaining the large number of movable and immovable elements of heritage keeps growing. Enforced by economic restrictions, we find ourselves more often confronting a demarcation between what can be considered as 'major' and 'minor' heritage in decisions about priorities for heritage conservation. It is not the subject of this chapter to discuss the ethics of such a differentiation, but, here, by 'minor' heritage we mean locally significant heritage which has meaning for the people of a neighbourhood or city, but which may not be formally recognised by protection in law or for its artistic or economic/touristic value.

The case study of Martos is a typical example of a locally significant heritage asset: a monumental fountain in a small town of Andalusia. How do you preserve it with your limited financial resources, since the state has different priorities? The reply to 'how?' came from the answer to 'who?'. Who is interested? The answer to this second question is: the local community. It is they who, in losing that heritage, lose part of their identity. It is their memories that are related to it, it is they who embody the intangible heritage that brings particular value to that specific built heritage.

In the local community and the power of synergy, Diadrasis saw an opportunity to propose a sustainable conservation intervention, through our project workshop system. We work on the principles of the barter system of an exchange economy: the local community provided housing, food, technical support and the human power of the local craft school, while Diadrasis would conserve the cultural heritage of the community with a team of international experts and participants, offering, at the same time, international awareness of the place.

But for this synergy to work successfully, a careful design and implementation of the interaction with the local community is indispensable. We first mapped and classified the different types of individuals and groups, then reflected on their characteristics and potential as to the conservation and maintenance problems of the heritage in question and gradually involved them in all different stages of the project workshop.

Case study: the fountain *Fuente Nueva*

The case study, *Fuente Nueva* (the New Fountain), is a monumental fountain designed in the sixteenth century, the project of architect Francisco del Castillo. Made mainly of sandstone in the Mannerist style, it has a highly decorated façade and two frontal basins: the upper and smaller one for the people and the lower, as wide as the façade, for the animals (Figure 16.5). The fountain played a very important role in the economic and cultural life of the town until the beginning of the twentith century. It was used as the main water supply for the residents in general and for animals, particularly during the annual cattle market.

An interesting fact is that the fountain is no longer standing in its original location. In the archives we found the documentation for at least two translocations, following its initial construction. The original location was on a square next to the church in the heart of the city, while, from the 1970s, it has been translocated to the end of a long shaded avenue, in a semi square, between residential palaces.

Analysing the archive data and the present conservation status of the fountain, we categorised the main factors that are threatening the conservation and maintenance as:

- material decay
- lack of archive data
- decontextualisation
- neglect.

To deal with these issues we designed a six-week project-workshop on stone conservation and urban regeneration. The main aim of this project was to work not only on the problems related to the stone decay, but also on issues related to urban planning and public awareness. As per the educational strategy for our workshops, all tasks should to be completed by the end of the workshop and, for effective management of the project, the subjects were divided into weekly modules as follows.

Week 1: Context, concepts and general methodology

Introductory week, including a short presentation of the area's history and an overview of concepts and general practices in conservation projects.

Week 2: Survey week

Introducing instruments and the methodology for documentation and survey and the documentation as a tool for conservation. By the end of the week a complete survey of the architectural elements was produced, which became the working base for the following weeks.

Week 3: Heritage management and dissemination/diagnostics

During the first part of the week we focused on sustainability, developing dissemination strategies and opening our conservation project to the local community. The second part of the week was dedicated to diagnostics – that is, the analysis of single elements and the pathologies of stone masonry.

Weeks 4–5: Stone conservation fieldwork

These were fully dedicated to the hands-on conservation experience, with open visits for the public to our works. By the end of the fortnight the intervention was completed, in collaboration with the students of the local craft school.

Week 6: Planning future activities and dissemination

The final week focused on the regeneration project proposal. That meant planning the regeneration of the urban area surrounding the fountain, aiming, with this conservation project, at incorporating the fountain in the contemporary urban life of the city again. The urban regeneration project has been subsequently carried out by the local craft school and the municipality.

Research with the public, conservation with the community

From the identified threats, the two main issues that we wanted to work on together with the local community were decontextualisation and neglect. At the same time we wanted to collect additional information from personal stories and from archives, filling archival gaps of the last century. To achieve these objectives we developed different strategies for all three distinct but interrelated phases of our project: comprehension, planning, implementation. During the comprehension phase we worked with the wider public; during the planning and implementation with different groups of the community. This schematisation of the developed strategy makes it easy to comprehend and to implement in other cases.

Comprehension: the public

A first step in all conservation projects is the comprehension and analysis of the case study. Along with the archival research we designed two activities: personal interviews and an open call for pictures. The main aim for these activities was to engage the attention of the wider public and integrate their knowledge and memories in the interpretation of the archive data. At the same time this was the moment of 'introducing' ourselves, preparing and raising expectations in the population of Martos for the workshop. It was, therefore, indispensable that our team and our activities be introduced to the people by the local authorities so as to gain their trust.

The interviews

The main aim of the interviews, conducted among the residents of Martos, was to understand their relationship with the fountain. Was it significant for them? If yes, why? Did they have some personal stories to share with us, related to the fountain, from its original or its current location? The analysis of the results demonstrated significant variations, according to different age groups.

Age group 50+
- remember the original location
- remember moments when it used to be the centre of economic and social life
- do not like the new location

Age group 27–50
- remember the fountain in use, with open access (today there is a fence surrounding it)
- are sceptical about any possibility of improvement of Martos's cultural heritage
- believe that the fountain must be incorporated again in the cultural life of Martos

Age group 18–26
- mainly local workers and university students
- participants of the *botellón*[1] (drinking evenings in the nearby park)
- university students: some concern for the damage to the fountain due to the *botellón*

Under 18 years old
- know the fountain from school visits organised by the Municipality's Cultural Heritage service
- very detached from the *Fuente Nueva*.

Call for pictures

Apart from typical sources of documentation, like photographic and graphic archives, with systematic documentation of the monument in question, there are also indirect sources of documentation, like personal pictures. Quite often, in personal pictures with the monument as background, you can collect very significant data for your research. So, through the local radio and the press, we organised an open call for collecting these personal images. In total we collected 37 pictures, which gave us valuable data on social activities, locations, dates and the state of the fountain's conservation, covering a time span of c. 80 years.

Planning and implementation

As interaction with the population of Martos kept growing steadily towards planning and implementation, we gradually passed from the wider public to the citizens. We do not distinguish planning from implementation because, when it comes to interaction with the local community, the success of all activities depended on the continuous connection between them. This means that the following actions were planned and executed with specific citizen groups, the local authorities, the craft school and the sensitised citizens focusing on specific problems to solve.

The local authorities

As introduced earlier, the key to success for this endeavour was the synergy between Diadrasis and the local authorities. This collaboration was the foundation of the entire project. It was primarily the passport for gaining the trust of the local population, a trust that could simply not have been gained during the six weeks of the project workshop by a team of foreigners, not speaking Spanish.

On practical issues, they provided not only the hospitality, but also mainly the constant technical support, offering staff from the various departments of the municipality:

- Urban Works and Maintenance department: construction of the scaffolding and provisional house of the worksite. Provision of a crane and technical support when, unexpectedly, the need came up to remove the central pinnacle and weather vane during the works
- IT: provision of IT room with eight computers and network support
- Cultural services: support before, during and after the project in communication with the citizens, provision of spaces for storage and exhibition.

Lastly, as the local authorities were also very interested in promoting the work done by an international team in their city, they constantly supported our work with the dissemination of the event with press, radio and TV coverage.

The local craft school

In Spain there is an established system of training workers for heritage craft works in the so called *Escuela Taller* or School Workshop. The *Escuela Taller* two-year training system is based on the combination of study and work (apprenticeship) and is part of the strategy developed by the *Junta de Andalucía* to combat unemployment in the region. Students in these schools are between 16 and 25 years of age and come from difficult socio-economic backgrounds. Collaborating with them, involving seven students under the supervision of one of their instructors, was an opportunity not to be missed.

This collaboration was yet another proof of interaction in action; when carefully planned it can give optimum results for all participating parties. The students of the school were initially trained by members of Diadrasis's staff in basic principles of conservation of monuments. They were then integrated into the working groups of our participants. On the worksite, their technical abilities and familiarisation with materials and tools became a unique educational tool for our heritage experts. In conclusion, not only did they actively contribute in the two weeks of conservation intervention of our workshop but, being the ones that will be taking care of the monument after our departure, they also ensured the sustainability of the intervention by indispensable periodical maintenance.

The enhanced awareness of citizens

During the six weeks of the actual project, great attention was given to the integration and interaction with that part of the population which, after the activities of the comprehension phase, became highly interested in the project workshop. Overcoming the language barrier, a large number of the people were curious about our team and our work. At first, they were curious about who we were, where we came from and what our plans for the fountain were. But, at the same time, they were also very willing to contribute, either by sharing memories or by sharing their opinion for a better future use and life of the fountain. In order, therefore, to keep their interest alive and to get the best out of them, both for our decision-making and for the sustainability of the project, we designed three different types of activities.

The exhibition

During the comprehension phase, with the call for pictures, we collected significant chronological images of the *Fuente Nueva*, depicting it in its original place and, after the two translocations, in its other two settlements. As the images also showed the relationship between the community and the fountain, we thought of organising a small photographic exhibition called 'Through the Eyes of the Lion', at the Municipality's *Casa de la Cultura*, with all the 'then and now' pictures of the fountain, thus starting a thread of continuity (Figure 16.1).

The exhibition, held in the second week of our project, was designed as an acknowledgement of the contribution of the citizens that shared their personal archives, but also as a connection moment between workshop participants and citizens of Martos. We also included a small presentation of our team and scheduled works, and invited them to participate in the other activities designed for them too: the 'referendum' and the 'Open for restoration' events.

The referendum

At this exhibition we held a small referendum, collecting opinions on the new urban planning our participants were to propose to the local authorities after completion of the project. To this end, voting postcards were made available at the exhibition for people to write their wishes and aspirations for the fountain's future. The outcome of this voting was that the majority of the community was not happy with the present urban and landscape arrangement of the fountain and they wished to see it relocated to a new place in the centre of Martos.

From the numerous suggestions, we have summed up a list of the most indicative ones:

- move the fountain
- demolish the adjacent buildings
- make the water drinkable
- make a new modern sculpture
- install new or better lighting
- remove the surrounding fence
- restore the function
- provide an historical explanation panel
- install a crosswalk.

Figure 16.1 Exhibition, contributing local showing and explaining the story of his picture with his schoolmates © DIADRASIS

The results of this referendum were subsequently integrated in the proposal for urban regeneration which we submitted to the Municipality upon completion of the project (Figure 16.2).

'Open for restoration'

From small monuments to big-scale monumental complexes, it is increasingly common that the momentum of conservation and intervention attracts great attention from the public. Contrary to the old style of 'secrets of the laboratory', more and more restoration projects include organised visits, to attract more visitors and, in some cases, like the *Cattedral de Victoria* in Spain, to actually raise funding through the entrance fee (La Torre 2011).

In our case, as the main aim was to regain the interest of the citizens and to enhance the age groups that our survey showed were the most indifferent to their heritage, we scheduled daily 20-minute visits in the yard of our worksite. The 'Open for restoration' event, was scheduled twice per day, once in the morning and once in the afternoon, so as to give the possibility for more people to visit. For safety reasons, people were not allowed in the actual working area but, as the worksite was in the open air, they had the possibility to see all the explained restoration treatments. This attempt was very successful, bringing in both spontaneous visits by individuals and scheduled visits by groups like university students and local clubs (Figure 16.3).

Evaluating the methodology

It is very difficult to estimate the successes and the failures of a methodology, when there are no quantifiable results. One can, however, attempt to indicate the outcome for the local population by evaluating the results on the identified problems.

Decontextualisation

From the discussions during the visits and the responses to the referendum, it was obvious that there is a clear connection between the urban location and the exclusion of the fountain, both from everyday life and big social community events. This decontextalisation is one of the main reasons for the neglect from the younger generations.

The implementation of an international workshop on the fountain created a new milestone for the importance of the fountain in the cultural and social life of the town.

Also, upon completion of the workshop, and as a demonstration of the Municipality's will to apply our guidelines for the urban regeneration of the surrounding area, two modern constructions that were blocking visual connection of the fountain with the park were demolished.

Figure 16.2 Exhibition. the moment of the voting referendum © DIADRASIS

Figure 16.3 'Open for restoration' organised visit © DIADRASIS

Finally, from the combined historical research and the people's stories, the design for an information panel was created by the dissemination team of our project, making one of the ideas from the referendum instantly come true.

Neglect

The number of participating citizens in our activities shows a positive response. In this small city, attracting the attention of the locals is quite easy. Just at the opening of the exhibition we had more than 150 visitors and many more visited all through its duration, as attested by the participations in the 'referendum'. During the two weeks of the 'Open for restoration' event, 178 persons visited our worksite, not only from Martos but also from neighbouring cities.

Great attention was drawn from the media (Diadrasis 2015). Two radio stations, two newspapers and the regional television dedicated time and space in their news both to the project and the monument.

The participation of the *Escuella Taller* was very significant as it brought about a result related to a difficult target group that it was hard to reach. The students and their friends are actually those who participate in the *botellón*, a result of which is often the vandalism of the fountain. By taking part in the conservation of the fountain, they became 'ambassadors' of the importance of the fountain, transmitting it to their groups and friends, becoming active guards of their heritage (Figure 16.4).

Conclusions

We will conclude by going back to the initial question: can the engagement of the local community be the key for a sustainable model of conservation in cases of minor, local heritage? The Martos project has definitely been a promising case study, as the results show (Diadrasis 2012). The success of the project was due to the fact that the interaction with the local community was planned in advance to run throughout the entire project, from the preparation, to the project workshop and, finally, to the continuity of the project.

Seizing the opportunity of a strong partnership with local authorities, we had the possibility to discover this new participatory and enhancing model of action. The workshop itself and the multiple interactive activities, turned the curiosity of the wider public into interest. The final results? Active citizens taking care of the fountain after the completion of the workshop and our departure prove that local communities can be a valuable answer for sustainable conservation, especially for locally significant cultural heritage (Figure 16.5).

Acknowledgements

The workshop was organised with the consent of the *Junta de Andalucía*, under the patronage of ICCROM and in collaboration with the University of York, the University of Jaén, the City of Martos Council, Department of Culture, ADSUR, the IAPH and the IPCE.

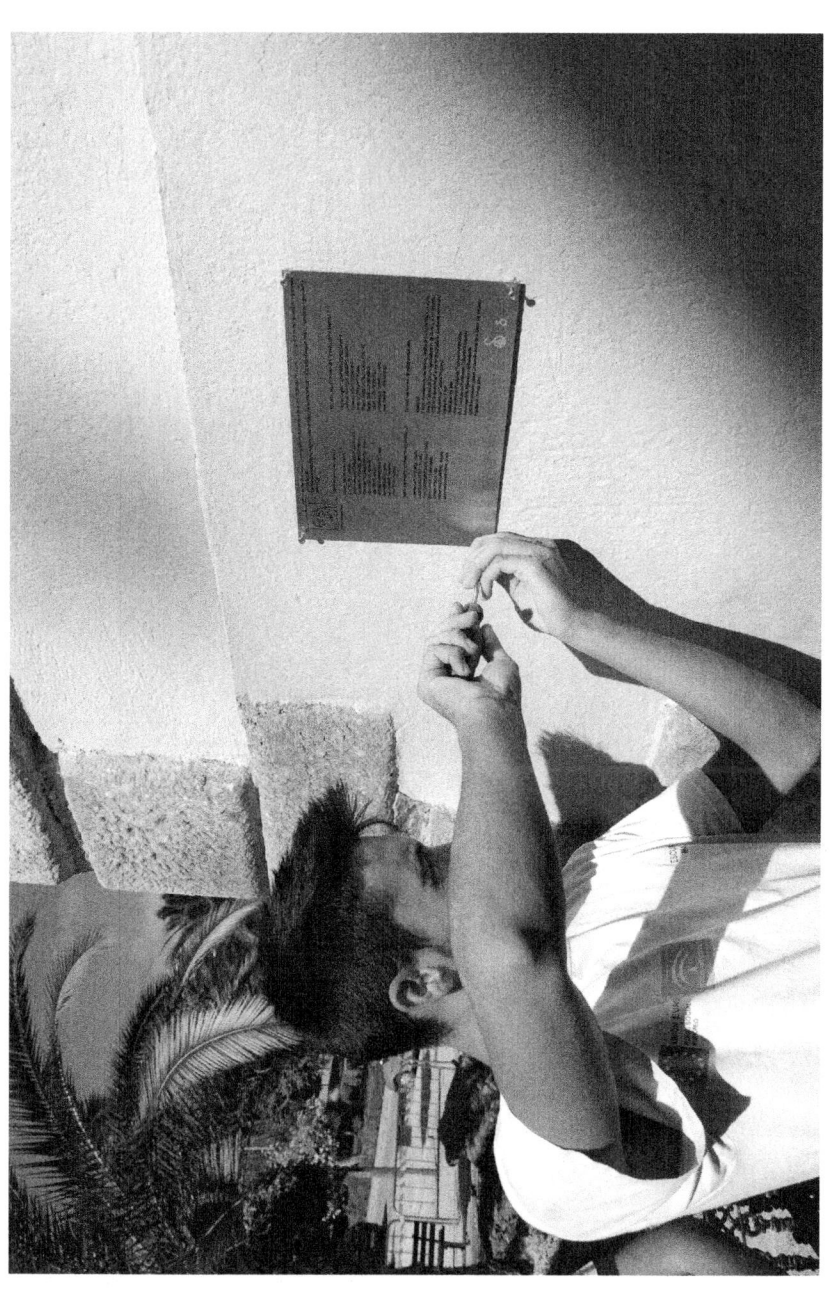

Figure 16.4 Student from the *Escuela Taller* fixing the commemorative plate marking the year of the intervention and the names of the participants © DIADRASIS

(11 participants+35 lecturers&specialists+7 craft students) x 42 days in Martos =

Figure 16.5 'Before and after' summary poster, including before and after pictures, the poster of the exhibition, the urban regeneration proposal and the interpretative panel design © DIADRASIS

Note

1 https://en.wikipedia.org/wiki/Botell%C3%B3n

References

Diadrasis 2011. The 'Romiri project' results [online]. Available at http://www.diadrasis.org/projects/romiri/results.html [accessed 7 September 2015]

Diadrasis 2012. The 'Martos project' results [online]. Available at http://www.diadrasis.org/projects/martos/results.html [accessed 7 September 2015]

Diadrasis 2015. DIADRASIS in the media [online]. Available at http://www.diadrasis.org/media.html [accessed 7 September 2015]

Georganas, I. 2013. 'The effects of the economic crisis on Greek heritage: a view from the private cultural sector'. *Journal of Eastern Mediterranean Archaeology and Heritage Studies* 1(3), 242–5

Hooper, J. 2010. 'Europe's heritage at risk from economic crisis'. *Guardian* 3 September [online]. Available at http://www.theguardian.com/culture/2010/dec/03/european-heritage-risk-economic-crisis [accessed 7 September 2015]

Inkei, P. 2010. 'The effects of the economic crisis on culture' [pdf]. Available at https://www.coe.int/t/dg4/cultureheritage/cwe/Effects_Inkei_EN.pdf [accessed 16 April 2015]

La Torre, P. 2011. 'The transformation processes of architecture through time: methodological and theoretical consequences in the project and in the restoration works' [pdf]. Available at http://www.diadrasis.org/dialogos/dialogos_001/edialogos_001-LATORRE.pdf [accessed 20 April 2015]

Muñoz Viñas, S. 2005. *Contemporary Theory of Conservation.* Oxford: Butterworth Heinemann

Tapini, L. and Gómez-Robles, L. 2013. *Analyzing the Project-based Workshop Model for Interactive Training in Conservation.* York Conservation: Fortieth Anniversary Conference, 6–7 July 2012. University of York

17 SPAB Maintenance Co-operatives

A move towards meaningful community participation?

Stella Jackson and Alaina Schmisseur

Abstract

The Society for the Protection of Ancient Buildings (SPAB) is unrivalled in its knowledge of the physical threats facing historic buildings and the techniques for minimising decay and damage. It also has a proven track record of successfully recruiting and motivating volunteers and of providing effective education to professionals, craftspeople and the general public. In 2006 the SPAB developed the acclaimed Faith in Maintenance (FiM) project, the success of which was recognised by a highly prestigious Europa Nostra Award in 2010 in the Education, Training and Awareness category. Education and training were the key elements of the FiM project, resulting in 'traditional', expert-led engagement via knowledge transfer. However, evaluation of the FiM project suggested that additional or alternative forms of engagement were required, and this has resulted in the new Maintenance Co-operatives Project. This three-year project brings together groups of people who care for places of worship and encourages them to work together to tackle the problems of maintenance and repair. This clear step towards participative practice and public engagement in heritage conservation resonates well with new international principles for capacity building. In relation to Sherry Arnstein's 'ladder of participation', however, do Maintenance Co-operatives push local volunteers up to Citizen Power level at the top of the ladder or do they remain somewhere in the middle at a more tokenistic level?

Introduction to the SPAB

The Society for the Protection of Ancient Buildings was founded in 1877 by William Morris in an effort to protect the historic fabric of ancient buildings against the threat of over-restoration. In 1880, the SPAB stood at 372 members and, over the intervening 135 years, membership has grown to over 8,000 (SPAB 2015). Although many of the members are conservation professionals, the SPAB membership is significant in that, from its beginning, it has not been limited to those with a formal background in architecture, historic preservation or conservation. The main focus of the SPAB has always been to raise awareness of the need to conserve building heritage, and this has been accomplished through a number of highly successful campaigns, many of which have been supported by volunteers who are drawn to the work due to their interest in history or in architectural conservation. This focus on ensuring that techniques and best practice for maintaining structural and historical integrity are accessible to anyone is integral to the mission of the organisation, and the SPAB has a long history of attracting, educating and supporting volunteers with a passion for buildings preservation.

Since its foundation, the SPAB has encouraged participative practice and public engagement through a variety of channels. The most visible and enduring of these are the publications and technical pamphlets, the first of which, 'Notes on the repair of ancient buildings', was issued in 1903. The first practical training courses were held in the early 1950s. By 1976 the need for technical training days in specific building materials was recognised with the first Lime Day training sessions, and in 1984 the first highly successful 'Introduction to the repair of old buildings' course was held in London. A formal telephone advisory service has also been provided to those with queries concerning the care of buildings and specific conservation issues free of charge since 1983 (ibid.).

The SPAB and places of worship

The Society for the Protection of Ancient Buildings has held a particular interest in the preservation of historic places of worship since the days of William Morris and the anti-scrape movement, and employs church caseworkers to assist parishes in adapting their buildings to modern usage while maintaining historic integrity (ibid.). In 2003, a series of church warden training days were instituted to further support wardens (who provide voluntary physical care for the church) in their efforts to identify maintenance and conservation issues and to better work with their architects and builders (Heritage Link 2003).

In 2006, the SPAB received a Heritage Lottery Fund (HLF) grant to institute the FiM project. This project was developed as part of the SPAB's effort to spread the message of good maintenance and consisted of one-day training courses held around the country and intended to support volunteers who are responsible for looking after places of worship, as well as a range of online guidance and advice. The FiM project was highly successful and ran for a total of five years from 2007, reaching a total of 4,585 attendees (Crofts and Minnis 2012); its success was recognised by a highly prestigious Europa Nostra Award in 2010 in the Education, Training and Awareness category (SPAB n.d.).

The Maintenance Co-operatives Project

The Maintenance Co-operatives Project (MCP) is the successor to the FiM project, and is funded for three years from autumn 2013 by the UK's HLF. It is a ground-breaking initiative to connect, encourage and support the dedicated volunteers across the country who are responsible for the upkeep of historic places of worship, many of which are in desperate need of repair and rely on the help of local supporters. The MCP has a similar focus on capacity building to ensure that places of worship are sustainably maintained, and is helping volunteers with this vital work by:

- raising awareness of good conservation practice
- promoting the benefits and techniques of timely maintenance
- increasing the number of volunteers actively caring for places of worship
- providing free training, support and resources
- facilitating the sharing of skills and knowledge
- making the most of opportunities to pool resources.

The new ICOMOS principles for capacity building (2013) suggest that education and training should be seen in the more general framework of capacity building and define education as the process of imparting knowledge and training as something designed to share skills along with a more specific knowledge and understanding (ICOMOS 2013: 1). Education and training were the key elements of the FiM project, resulting in 'traditional', expert-led engagement via knowledge transfer. However, evaluation of the project suggested that additional or alternative forms of engagement were required, although 81.12 per cent of FiM participants agreed that the course increased their enjoyment of their voluntary role and 83.92 per cent reported that they felt more confident about identifying maintenance issues or problems (Crofts and Minnis 2012). The new MCP aims to bring together groups of people who care for places of worship, encouraging them to work together to tackle and to better understand the problems of maintenance and repair. This fits well with the new ideas about capacity building as a people-centred process by which individuals or groups develop the abilities to perform functions, solve problems and set and achieve objectives (ICOMOS 2013: 2–3).

The MCP is achieving this through creating a series of locally based networks of volunteers called Local Maintenance Co-operatives, who share good maintenance practice. A main aim of the project is to build capacity so volunteers can more confidently carry out essential preventative maintenance themselves. Following an initial training day, Co-op groups carry out their own maintenance audits and develop their own maintenance schedules. The project is also offering a range of free training and skills development opportunities to those responsible for or interested in the care of places of worship. In addition to a training programme for each region, Co-operatives also benefit from a tailored training programme which allows them to direct their own learning based around local need and demand. In Lincoln, for example, volunteers have asked for training on open churches and tourism, whereas a group near Boston have asked for technical advice on topics such as dealing with damp and re-pointing.

The use of the term 'Co-operative' refers to the volunteers, not the buildings, and the purpose of a Co-operative is to share and nurture the expertise of its members so that they can support each other in maintaining their local places of worship. Although Co-operatives are initially composed of current church wardens and stewards, membership is open to anyone who supports the purpose of the group and participates in its activities. Once established with the assistance of the MCP team, Co-operatives are self-managed groups – they are not part of a charity with a participation policy and do not work directly under the leadership of expert-led conservation teams.

Supporting new volunteers

Social connections are central to participation as the desire to make social connections, meet new people and combat isolation or loneliness can trigger involvement in a range of voluntary activities (Brodie *et al*. 2011: 70), and is a key idea behind the MCP. The relationships that are built in groups are also a crucial sustaining factor in people's participation (ibid.). Membership of Co-operatives is intended to be enhanced by the inclusion of new/additional volunteers from the local community, especially non-worshippers who may value their nearby place of worship as a key element in their cultural landscapes and thus wish to see it retained. Because they are not part of the congregation, however, they may feel that they cannot get involved with its upkeep or even go inside. This issue of cognitive ownership has always been a problem for religious built heritage (Boyd 2012: 2). The term 'cognitive ownership', coined by Boyd and Cotter in 1996, reflects the links that people make with heritage places, which are defined by the intellectual, conceptual and/or spiritual meanings that become attached to them (Boyd 2012: 1, 4). Therefore, although most members of a local community will usually value their nearby place of worship and may wish to participate in its care, there is often a metaphorical barrier that exists for those who are not worshippers as they do not consider themselves eligible to do so. A sense of place and ownership, however, is often a strong outcome of engaging and directly involving people in the care of their local heritage and can in turn foster a desire to protect assets for future generations (Brown 2014: 79).

Climbing the ladder

While the Maintenance Co-operatives Project tenets would appear to align very well with the ICOMOS Principles for Capacity Buildings, in relation to models of community engagement and participation (such as that of Sherry Arnstein (1969)), can Local Maintenance Co-operatives be said to be fully engaged and participative? To examine this, first is a brief outline of Arnstein's ladder of participation and other similar models, including that of the HLF, the main funder of the MCP.

In her seminal article, Sherry Arnstein outlined the tension that often exists between the 'haves' and the 'have-nots' in relation to decision-making and argued that participation without the redistribution of power is empty (ibid.: 216). Although this is written in relation to the participation of citizens in government decisions about where and how they live, the basic principles can be used to assess all forms of participation, so it is a useful tool for considering the roles of volunteers within heritage conservation. The focus of Arnstein's attention was the poor practice she observed in her own work and the work of others seeking the meaningful engagement of existing and potentially new participants (Barber 2007: 23). In her analysis of participation and non-participation she used a typology of eight levels arranged in a ladder pattern, with groups effectively trying to climb up to the top rung (Figure 17.1, Arnstein 1969: 217).

The ladder juxtaposes powerless citizens with the powerful in order to highlight the fundamental divisions between them. Categories begin with those of a non-participatory nature at the bottom of the ladder and finish with increasing degrees of citizen control at the top. The highest rung of 'citizen control' is defined as occurring when the 'have-not' citizens obtain the majority of decision-making seats, or full managerial power (ibid.). If citizen control under Arnstein's typology is defined as participants governing a programme – being in full charge of its management, and being able to negotiate the conditions under which 'outsiders' may change it (ibid.: 223) – then the Co-operatives project could clearly be categorised as such, and this is a key aim of the project.

Participatory democratic theory tends to assume that positive outcomes arise from participatory involvement, with benefits for both individuals and society at large (Head 2007: 448; see also Neal and Roskams 2013: 81); and advocates of strong and effective forms of participation, such as Arnstein, argue that that citizen control is the level of participation to which all should strive, with the lower rungs often being defined as 'weak' (Head 2007: 444; see also Hart 1997; Walters *et al.* 2002; Ross *et al.* 2002; Bishop and Davis 2002). Roger Hart (1997), for example, refined and created a new version of the ladder in relation to youth participation, but still implies that the lower rungs are of little value, which does not take into account that young people maybe just are not ready to climb the ladder (Barber 2007: 26). The International Association for Public Participation define five levels of participation, which range from informing to empowering, with the goal for empowerment being to place final decision-making in the hands of the public (IAPP 2015).

Arnstein's model has received both criticism and refinement since it was first published in 1969 (see, e.g., Barber 2007; Bishop and Davis 2002; Hart 1997; Head 2007; Ross *et al.* 2002; Treseder 1997; Walters *et al.* 2002). The main criticism is that it is rather stereotypical and over-simplistic, presenting

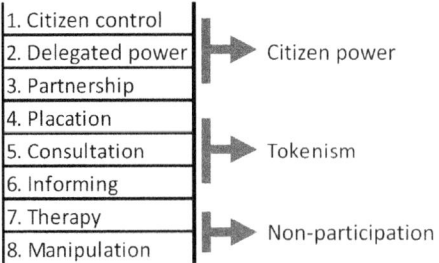

Figure 17.1 Ladder of participation © Authors, adapted from Arnstein 1969: 219

the different stages without context and making assumptions about progression from one stage to another without taking account of the dynamic and unpredictable nature of participation (Barber 2007: 24–5). Although Arnstein does acknowledge that things are much more complex in the real word, her model is most appropriately applied in promoting democratic involvement in a local government context (ibid.: 30), as these models all tend to be written in relation to government decision-making, with communities campaigning for a role within this which they hope will assure better outcomes for their local area. The Maintenance Co-operatives Project, therefore, does not fit as easily within this model, especially as anything other than basic maintenance work would typically need to be undertaken in partnership with the architect or surveyor for each place of worship and will often require an application for consent.

However, the MCP is funded by the HLF and community participation is a key aspect of their grants programmes. In their guidance on this and how to include it in projects, they also outline five levels at which people could participate in a project (HLF 2010: 5), adapted from a version of Arnstein's ladder of participation by David Wilcox (ibid.: 2; Wilcox 1994). The HLF levels of participation are as follows:

1. Informing: telling people about the project
2. Consulting: giving people choices about the project and asking for their views
3. Deciding together: opportunities for people to make decisions and influence direction
4. Acting together: opportunities for people to develop and deliver the project
5. Supporting others to take the lead: empowering people to have ownership of the project, to make final decisions and to deliver activities with some independence.

The HLF guidance suggest that Informing and Consulting, stages one and two, are important first steps in a project, but are passive ways of involving people and will not fully meet their participation aim, which requires something more substantial (HLF 2010: 5). The overall aims of the MCP are effectively those of level five in the HLF model, which is similar to the Citizen Power rung of Arnstein's ladder and no surprise given that the project is HLF funded. However, there is an assumption within all of these models that full empowerment and project ownership is significantly better and more purposeful. This assumes that all people desire to be empowered and that those who do not or are not are 'missing out' in some way. This is not always the case in volunteering groups, as not everyone wants to be involved in decision-making and the management of a project (as will be discussed below). This is an issue that needs some critical consideration and is perhaps something to be challenged if we are to fully understand and provide benefits for volunteers.

Volunteers

There is, arguably, an assumption that empowerment is something that everyone wants to achieve. However, volunteering can be a core part of people's lives or something that they do once in a while (Brodie *et al.* 2011: 5), and not all volunteers wish to be involved at such a committed level. People volunteer for a range of different reasons and vary widely in the commitment they are able to give (HLF 2013). An over-optimistic view of participation and volunteering can portray it as a universal panacea and it is important that we acknowledge its limitations and develop realistic expectations of what can be achieved (Brodie *et al.* 2011: 8). Volunteer motivations for participating in groups such as Local Maintenance Co-operatives are diverse and complex. The most popular reasons for volunteering are, according to research, to make a difference and/or because of strong feelings about a cause (Jochum *et al.* 2011: 28). As people age they spend more time volunteering; the most active volunteers, therefore, tend to be retired professionals who are often happy to carry out organisational and managerial roles, especially when they have taken early retirement from a similar role (Brown 2014: 79; Jochum *et al.* 2011: 27). In contrast, younger volunteers often take on roles to gain valuable experience for their CV, while others enjoy the benefits of volunteering but are time-poor. The perceived high workload of volunteers, along with a range of other commitments competing for their time such as families and careers, may mean that they do not wish to take on extra responsibility or time commitments, preferring to give a more flexible contribution (Brown 2014: 80; Holmes and Slater 2012: 852; Jochum *et al.* 2011: 28).

It is clear that many Co-operative volunteers would prefer to remain at the so-called 'tokenistic' level of participation, passively volunteering by attending training courses to further their knowledge and suggesting topics for advice, guidance and further training, but nothing more. However, research carried out by Brodie *et al.* (2011: 72) challenges assumptions that non-participation is a result of apathy, laziness or selfishness. They found that people juggle many competing demands for their time and attention with their priorities varying according to personal circumstances and life stage, which has implications for how and when they might become actively involved (ibid.). Opportunities to participate, therefore, need to respect the needs, motivations and expectations of potential volunteers, and the MCP has tried to do this by taking into account ideas that came out of the FiM evaluation and feedback from event attendees.

Moving towards meaningful participation?

Despite concerns over the participation models outlined above, HLF guidance does suggest that participation is very much dependent on the project and different levels are appropriate in different circumstances (HLF 2010: 7). Thus, while the project has delivered a wide range of training and other events, progress has varied both within and across the project regions, with some groups preferring a more traditional programme of training that they can simply attend for interest, and others being very specific about their needs The notion that participation at the bottom rungs of the ladder is less effective than self-directed participation, therefore, may be flawed. Work by Treseder (1997), for example, shows that consultation may be far more desirable for people who may not feel comfortable or motivated to fully participate. A consultative approach may be far more appropriate and less tokenistic than is assumed: people still feel that they have made a real difference by participating in this way (Barber 2007: 30). In practice, therefore, participation may have a dynamic which is much more complex than models such as Arnstein's ladder of participation suggest. There is a need to fully explore the contextual nature of participation and the processes which influence engagement and dialogue (ibid.: 35). For the MCP, it might be far more appropriate to view informing and consulting as effective participation, especially in relation to the capacity-building aim of the project.

Case Study: North Manlake Co-operative, Scunthorpe, North Lincolnshire

The North Manlake Co-operative, covering an area just North of Scunthorpe, was one of the first to be developed by the MCP and began to meet early in 2014, although the churchwardens involved had already discussed how they might make it work before the project officer was even in post. They relish the idea of managing themselves and making their own decisions on priorities about which they need to speak to their architect in terms of conservation and repair projects; they are also very positive about the aim to build social capital in the area by developing and encouraging connections that enable them to draw on wider groups of both people and resources (Head 2007: 443). Co-op group meetings (Figure 17.2) are regularly attended by a wide number of people from the six member churches, and meetings are also attended by the recently formed Friends of All Saints Church in Winterton.

This group has been very specific about the training they would prefer and how they would like the group to run, and is already seeing the benefits of working together. They are sharing knowledge of local trades people, who are known to do good work, and collaborating together to undertake baseline condition surveys of the churches involved in the group. In relation to formal technical training topics, however, such as dealing with damp or understanding lime, the group is happy to attend these as they are arranged on a regional basis rather than requesting them specifically for the group. This approach differs from what the Lincoln City Co-operative has done, for which a very detailed training programme has been developed. A key element of volunteering is that people gain as well as give when they participate, since people's involvement is usually dependent on mutually beneficial activity. This can be as basic as enjoyment and having fun (Brodie *et al.* 2011: 70). A key aspect of the success of the North Manlake group, therefore, is that there is a real social and community aspect to it, with members enjoying the chance to get together and see each other. When asked what they would like to gain from joining the group, although many said maintenance skills and

Figure 17.2 North Manlake Maintenance Co-operative meeting © SPAB

support it was also clear that they also hoped to have fun as part of a social group (as Figure 17.3 indicates).

The Co-op members are also building social capital by sharing with the wider community skills and knowledge gained at what Arnstein would class as token-istic participation. Following a recent training event on caring for churchyards, as part of a meeting, which included a talk from Lincolnshire Wildlife Trust, a number of representatives from the churches involved have begun to think about creating areas of wildflower meadow in their churchyards. A representative from one of the churches, in fact, called the Lincolnshire Project Officer the next day to advise her that he had returned home after the meeting and already had three volunteers from the village ready to begin work, all of whom are keen gardeners but do not have gardens of their own. The churches involved are also using main-tenance skills learned as a group, which is resulting in a positive impact on the condition of the buildings. Although the administrative structures of the Anglican Church mean that they are not, and never can be the final decision-makers, for the majority of the work that is undertaken to repair the churches, having a better understanding of their building and confidence in its appropriate care and main-tenance is helping to safeguard these buildings for the future.

Conclusion

As the Maintenance Co-operatives Project team has discovered, what may be key to meaningful participation is the development of new ways for volun-teers to become involved and the generation of a wider range of volunteer roles which can more easily take into account the different levels of participa-tion to which volunteers wish to commit (Brown 2014: 80). Participation is dynamic and opportunities need to be flexible (Brodie *et al.* 2011: 8). The Maintenance Co-operatives Project does allow this flexibility, and those who prefer more 'traditional', knowledge-transfer participation may still be involved while those who wish to go further may set themselves up as a Maintenance Co-operative.

The ways in which people volunteer and the circumstances under which participation takes place are complex, and we need to reflect and respond to this (ibid.: 69). Having the understanding that participation occurs at a range of levels, all of which are appropriate depending on the circumstances and those involved, indicates that the MCP should be able to meet its aims and build capacity at the local level. This may mean that volunteers are better able to maintain their places of worship through attending a range of training sessions and events; or it might mean that they develop a self-supporting group, the members of which help each other with the maintenance of a number of places of worship, sharing ideas, resources, equipment and volunteer time between them. Neither of these approaches is necessarily better or worse, it is simply a different way of ensuring that our historic places of worship are maintained and conserved for the future.

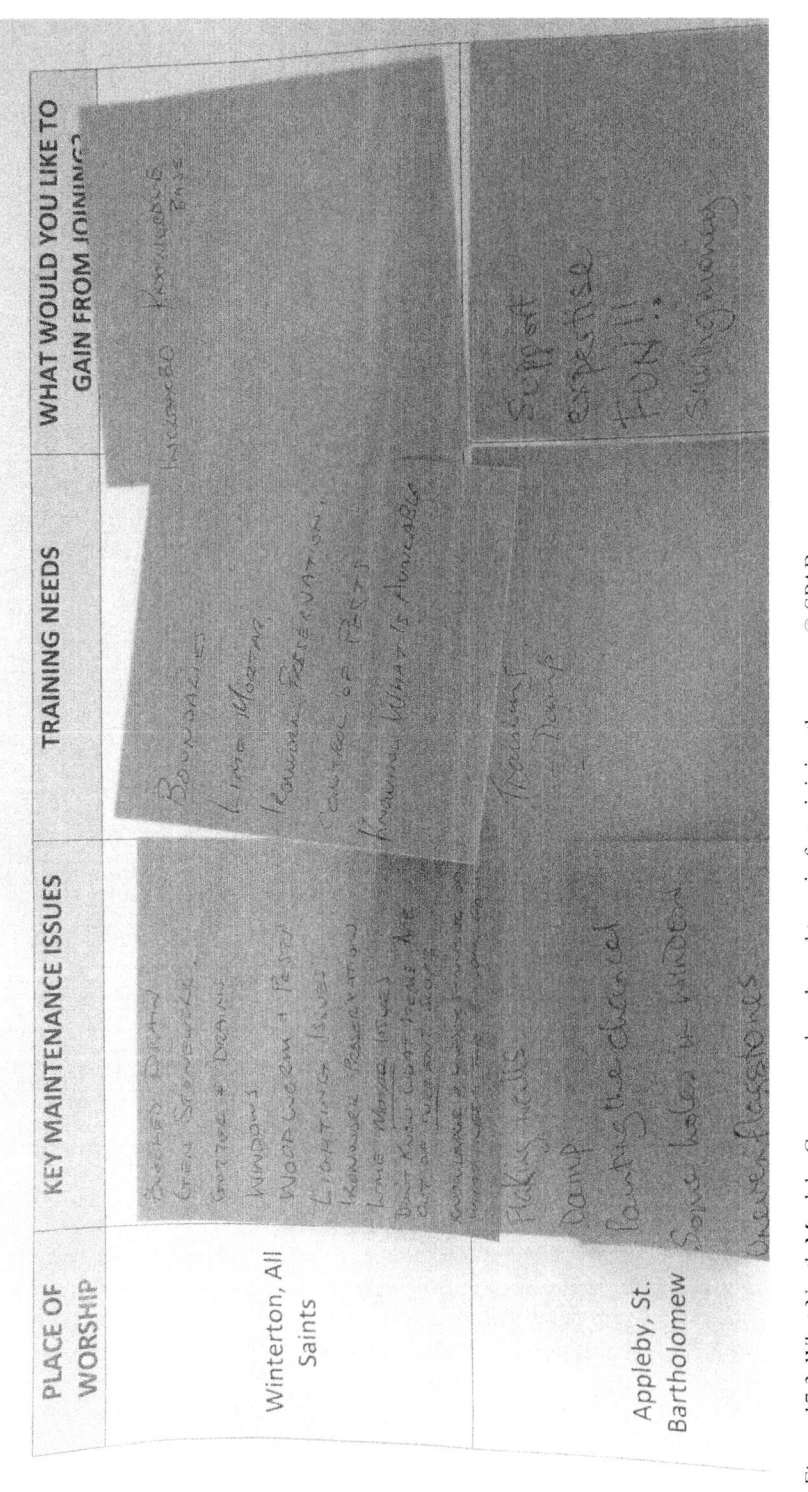

Figure 17.3 What North Manlake Co-op members hoped to gain from joining the group © SPAB

References

Arnstein, S.R. 1969. 'A ladder of participation'. *Journal of the American Institute of Planners* 35(4), 216–24

Barber, T. 2007. 'Young people and civic participation: a conceptual review'. *Youth and Policy* 96 (Summer 2007), 19–39

Bishop and Davis 2002. Mapping Public Participation in Policy Choices. *Australian Journal of Public Administration*, 61(1), pp.14–29

Boyd, W.E. 2012. '"A frame to hang clouds on": cognitive ownership, landscape, and heritage Management'. In R. Skeates, C. McDavid and J. Carman (eds), *The Oxford Handbook of Public Archaeology*. Oxford: Oxford University Press

Brodie, E., Hughes, T., Jochum, V., Miller, S., Ockenden, N. and Warburton, D. 2011. *Pathways Through Participation*. London: NCVO

Brown, L. 2014. 'Community archaeology and volunteers'. In S. Jackson, R. Lennox, C. Neal, S. Roskams, J. Hearle and L. Brown, 'Engaging communities in the "Big Society": What impact is the Localism agenda having on community archaeology?'. *The Historic Environment Policy and Practice* 5(1), 74–89

Crofts, S. and Minnis, K. 2012. *Faith in Maintenance Project Review*. Available at http://www.spabfim.org.uk/data/files/pages/faith_in_maintenance_project_review.pdf [accessed 8 August 2015]

Hart, R. 1997. *Children's Participation: The Theory and Practice of Involving Young Citizens in Community Development and Environmental Care*. London: Earthscan

Head, B.W. 2007. 'Community engagement: participation on whose terms?'. *Australian Journal of Political Science* 42(3), 441–54

Heritage Link 2003. *Volunteers and the Historic Environment: A Report by Heritage Link*. December

HLF (Heritage Lottery Fund) 2010. *Thinking about Community Participation*. London: Heritage Lottery Fund

HLF 2013. *Volunteering: Good Practice Guide*. London: Heritage Lottery Fund

Holmes, K. and Slater, A. 2012. 'Patterns of voluntary participation in membership associations: a study of UK heritage supporter groups'. *Nonprofit and Voluntary Sector Quarterly* 41(5), 850–69

IAPP 2015. *International Association for Public Participation: Core Values*. Available at http://www.iap2.org/?page=A4 [accessed 14 October 2015]

ICOMOS-CIF 2013. *Principles for Capacity Building through Education and Training in Safeguarding and Integrated Conservation of Cultural Heritage*. Available at http://cif.icomos.org/pdf_docs/CIF%20Meetings/Guidelines/ICOMOS_CIF_Principles Capacity_EN_20130930.pdf [accessed 14 October 2015]

Jochum, V., Brodie, E., Bhati, N. and Wilding, K. 2011. *Participation: Trends, Facts and Figures*. London: NCVO

Neal, C. and Roskams, S. 2013. 'Authority and community: reflections on archaeological practice at Heslington East, York'. *The Historic Environment: Policy and Practice* 4(2), 139–55

Ross, H., Buchy, M. and Proctor, W. 2002. 'Laying down the ladder: a typology of public participation in Australian natural resource management'. *Australian Journal of Environmental Management* 9(4), 205–17

SPAB (Society for the Protection of Ancient Buildings) 2015. *History of the SPAB* [online]. Available at http://www.spab.org.uk/what-is-spab-/history-of-the-spab/ [accessed 6 August 2015]

SPAB *Faith in Maintenance, About Us* n.d. Available at http://spabfim.org.uk/pages/europa_nostra.html [accessed 17 August 2015]

Treseder, P., 1997. *Empowering Children and Young People: Training Manual*. London: Save the Children

Walters, L.C., Aydelotte, J. and Miller, J. 2002. 'Putting more public into policy analysis'. *Public Administration Review* 60(4), 349–59

Wilcox, D. 1994. *Guide to Effective Participation*. Available at http://www.partnerships.org.uk/guide/ [accessed 14 February 2014]

18 Maintaining treasures on Earth

Supporting volunteers to care for places of worship

Henry Russell and Philip Leverton

Introduction

This chapter will examine the way in which those who care for our places of worship on a voluntary basis can be engaged by the conservation sector. It will look primarily at the Church of England, since this provides a good illustration of the importance of local ownership and local leadership in the conservation of heritage assets.

The primary responsibility for the care of churches falls on volunteer church members who, through parochial church councils, are managing all aspects of their church's life. In many instances, they are expected to care for major historic buildings without experience. In seeking to do so, they are faced with the challenge of respecting sound practice concerning the maintenance, repair and presentation of buildings at the same time as ensuring their viability both as settings for worship in the twenty-first century and as a venue for an increasing range of other functions.

This challenge was graphically described in the *Guardian* (Dakers 2011) thus:

> Fabric is an ecclesiastical obsession. While the maintenance of its holy ground can be delegated to a couple of old (voluntary) codgers, attending to its grass, graves and hedges, the stone work, organ, glass, roof and heating have to be undertaken professionally, and frequently; most of God's houses are of ancient origin – and under-insulated.

Support has been provided to local parishes by a variety of means at diocesan level and at national level in the Church of England. Other support schemes have been provided from outside the church by, most notably, the Society for the Protection of Ancient Buildings (SPAB).

The authors will examine the issues facing local parishes, whether the current levels of support are adequate and make recommendations for filling the gaps in support. It will also examine how professional conservation education can give practitioners the skills to support church communities. How do we prepare emerging newly qualified practitioners to be able to act increasingly as facilitators alongside their professional skills? How do we develop the ability to 'inform about informed conservation' in a professional environment that presently is being driven by the triple imperatives of context/significance, sustainability and localism, where the ability to use a wider range of skills and competencies to manage heritage assets is of growing importance?

Enhancing the effectiveness of available tools

Despite a growing range of training, advice and toolkits, we must ask ourselves whether it is hitting the larger part of the target audience or just a limited proportion of it. There are well-documented examples of how successful use has been made of available resources.

The SPAB developed the Faith in Maintenance (FiM) programme in 2007 and this ran for five years until 2011. It was offered to all denominations as a one-day course taken to individual regional church groups and targeted at the volunteers who are responsible for the care and maintenance of church buildings.

Over half of the Anglican dioceses ran courses, but take-up by other denominations and faiths was limited. Only five courses were held for the Roman Catholic church, and three of these were for one diocese, the Clifton Diocese (Crofts and Minnis 2012: 4). The greatest coverage was, not unexpectedly, in the Anglican church (Church of England and the Church in Wales) and courses were held in all Church of England dioceses (ibid.: 4–6). Table 18.1 demonstrates the number of courses run on a regional basis across all faiths and denominations.

Crofts and Minnis (ibid.: 23) concluded that the programme had a beneficial effect on conserving ecclesiastical heritage and supporting volunteers in doing this. Their report was followed by an independent assessment of the project, which concurred that there had been increased awareness of the need for maintenance, an improvement in the skills and motivation of volunteers and better maintenance regimes for places of worship (Goddard 2012: 23). The SPAB recognised the need to provide a legacy for the FiM programme. Continuing

Table 18.1 Faith in Maintenance

Courses by region (based on HLF regions)	Number of courses	Delegates	Per cent split of delegates by region	Average per course
London	12	350	8	29
South East England	24	762	17	32
South West	23	675	15	29
West Midlands	14	446	10	32
East Midlands	19	593	13	31
East	17	537	12	32
North West	10	304	7	30
North East	7	183	4	26
Yorkshire and Humber	12	375	8	31
Wales	10	323	7	32
Isle of Man	2	37	1	19

Source: Crofts and Minnis 2012: 6

support for the programme is provided by a handbook (SPAB 2008) and the FiM website (SPAB 2015) and by the SPAB's successor Maintenance Cooperatives Project (MCP, discussed in Chapter 17).

The Church of England recognises the need to support volunteers, and recent years have seen provision and support of a more proactive nature. This takes three forms:

1. *Simplifying processes:* The Faculty Simplification Project commenced in 2012 and will be complete when the Faculty Jurisdiction Rules 2015 (FJR) take effect in 2016. The objective has been to simplify the procedure of applying for a faculty,[1] while maintaining standards of control. Some works such as repairs identified in quinquennial inspection reports will no longer require a faculty, but will still need to be considered by the DAC (FJR 2015, Schedule 1). Provision has also been made for online submission of faculty petitions (applications to carry out works to a church) with simplified forms (Church Buildings Council 2015a).

2. *Improved guidance documents on Church of England websites*: A suite of guidance documents, prepared by the Church Buildings Council (CBC) (Church Care 2012), cover statements of significance and need, for which templates are provided,[2] conservation management plans and technical aspects of repair and maintenance. Individual dioceses also produce good guidance for their parishes and an example of one such is Exeter Diocese (Diocese of Exeter 2015), which provides advice on procedure and technical matters.

   The CBC provides advice to parishes about using church buildings for a wider range of purposes to help them become more economically sustainable under its Open and Sustainable programme. Council officers provide strategic support and the Church Care website is a source of useful case studies and support (Church Buildings Council 2015b).

3. *Providing targeted support to churches undertaking major projects*: In the last few years the CBC has been giving support to some churches through the production of conservation management plans by the CBC project officer. Examples of this are the CMPs for St German's Cornwall and St Mary the Virgin, Oxford (Church Buildings Council 2015c). CBC members are also providing support to major projects such as Bath Abbey, St Mary Redcliffe in Bristol and St John Waterloo in Southwark. This is often done through CBC members sitting on parish project advisory boards. There is a potential risk of conflict of interest when the CBC is consulted formally on the project, and this is mitigated by the formal consultation being undertaken by CBC members and officers not directly involved in the project.

Capacity constraints in applying the tools: a case study

A substantial number of parishes within the Church of England (particularly in rural areas) have initiatives for developing proposals for using church space imaginatively to meet both liturgical and community needs for the twenty-first century: these may be in the hands of a small number of individuals with the necessary time and energy to contribute. Human and financial resources are frequently very limited, the demographic of parishioners may be aging, the parish may be part of a group or benefice where clergy and services are shared between multiple churches (none of which will be exempt from the quinquennial inspection), and the demands that this will bring for addressing general maintenance and repair. In short, the bucolic whimsy of the BBC's *The Vicar of Dibley* is giving way to the grittier reality of the same broadcaster's *Rev*.

For instance, consider the example of this parish situated in a rural area within the East of England. Its identity has been anonymised by agreement with interviewees. However, it is typical of the two-thirds of the parishes that responded to the survey undertaken for the Church of England's Archbishops' Council Faculty Simplification Group (FSG) (ACFSG 2012). Its church, containing elements dating back to the 12th century, is listed Grade 2*. It forms part of a grouping of five parishes with responsibility for six churches between them, overseen by a single Priest-in-Charge.

> The parish has a population of 130, but no local facilities; there is no village hall, the only pub closed in the mid-1960s and the shop and post office a few years later. The Parochial Church Council (PCC), in consultation with the Priest-in-Charge, recently wished to undertake two projects. The first proposal was to convert the north porch (currently used for storage, and never to gain either access to or egress from the church other than in emergency) into a small kitchen. The second was to reorder the nave by removing the fixed pews in order to provide a space that would be capable of flexible use for both worship and community activities. These proposals were consistent with the third most common reason for applying for a Faculty, which were mentioned by 14 per cent of the respondents to the FSG's survey (the most frequently identified were, predictably, roof and fabric repairs, at 23 per cent and 20 per cent respectively).
>
> Responsibility for developing these proposals and negotiating with the Diocesan Advisory Committee for the Care of Churches (DAC) in pursuit of the necessary Faculty was placed in the hands of a working group of three members. However, the working group advised the PCC not to proceed with this scheme for the present because it regarded the process – obtaining the necessary consents, raising the funding and overcoming opposition or indifference on the part of some members of the congregation (for whom the church is a 'hallowed space' that should not be changed) – as too complicated and time consuming. This was despite the implementation of the simplified Faculty regime after the project had commenced.
>
>   Information provided to the authors in interviews
>   with a churchwarden, April and July 2014

From this information, it seems that the parish decision has less to do with the significance of the church as a building and the impact of the proposed changes on the fabric (important as these are) than with the self-perceived lack of capacity on the part of lay volunteers (Lukka *et al.* (2003) have shown that this phenomenon is not unique to the Church of England). Their situation has not been improved by the retirement of the Priest-in-Charge in April 2014, with the inevitable interregnum before a successor is in place, plus the prospect of the group being enlarged by the inclusion of three more parishes.

This parish in not atypical in terms of scale and amenities; it falls within the model population size of 100–499 for the county concerned. Consequently, it is reasonable to suggest that the shortage of local capacity may not be atypical either. Certainly, the general situation here appears to reflect the comment included in Appendix 2 to the FSG's report that its survey found that 'many parishes feel that the present process was difficult' (ACFSG 2012: Appendix 2, 9). Clearly, the simplification introduced following adoption of the FSG's recommendations should have alleviated matters, but it is insightful that the conclusion to Appendix 2 refers somewhat ominously to those parishes that had been through an application for a Faculty as having 'survived the process' (ibid.).

Facing the challenge of how to enable capacity building

So how is this challenge to be addressed? How do we develop the necessary skills in practitioners to help develop capacity in what the FSG's report describes as the 'unpaid volunteers whose services in many parishes are a scarce and over-stretched resource' (ACFSG 2012: 6) in engaging with a Faculty system (whether simplified or not). In a broader context, this also has a bearing on the feasibility of the UK government's localism agenda – despite the arguments that this agenda is over-rich in rhetoric but poor in terms of real engagement. Within the context of the parish represented in the short case study above, the Centre for Rural Economy at Newcastle University (2014: 9) has argued that if power and responsibility are devolved to rural communities without adequate resources and support this is 'a recipe for a two-speed countryside', represented by those with and those without skills, assets and institutional capacity.

Enabling and capacity building are critical skills for professionals to learn, and are not a skill-set that is given extensive coverage in the curriculum of most current conservation programmes within Higher Education. Alongside specialised content relating to the philosophical, historical and technical aspects of conservation, generally (and properly), to one extent or another there will be attention paid to developing students' capabilities in what may be termed 'standard' transferable skills such as team working, time management and making presentations. Most universities and colleges would require this as a necessary component of any programme being brought forward for approval within the institution.

However, the ability for emerging practitioners to 'inform about informed conservation' also depends upon a higher level set of transferable skills that may be used to close what Sayce and Farren-Bradley (2011: 32) characterise as the 'perceptual gap between professions and public' requiring a challenge to the latter. In this context, Sayce and Farren-Bradley (2011: 23–4) suggest that practitioners have 'to rethink their assumption of "working for" into one based on a notion of "working with"', and draw attention to research into the nature of professionalism that identifies 'the role of the professional as "mediator" between the state and the individual'. We may go further here and suggest that mediation skills would be valuable in resolving differences of opinion between opponents and promoters of a scheme.

Given the growing emphasis on the sustainability dimension to conservation, some of the literature that has emerged during the last decade on developing professional and community capacity with regard to sustainability applies equally well to heritage conservation and management with regard to the development of the higher level skills referred to above.

For example, echoing Sayce and Farren-Bradbury, Murray (2011: 204), in a wide-ranging review of skills required to promote sustainability, emphasises the importance of interpersonal competencies. In particular, this includes the ability to communicate, the capacity to work effectively with others and the awareness to provide good leadership in appropriate circumstances. English Heritage (now Historic England) (2013: 9) has argued that '[t]raining should not just focus on specialist skills but also on generic work-based skills, and Higher Education courses should consider teaching a broader range of skills'. However, Moore *et al.* (2015: 146) have warned that the latter are not broad generalities, applicable in any situation; rather they have to be refined and adapted to the particular circumstances that are being dealt with.

The UK government-commissioned *Egan Review* published in 2004 put forward an ambitious menu of sought-after generic skills required to enable what it titled as: inclusive visioning; project management; leadership; breakthrough thinking/brokerage; team/partnership working; making it happen given constraints; process management/change management (ODPM, 2004). A subsequent report, *Mind the Skills Gap*, by Arup, on behalf of the Academy for Sustainable Communities (2007), disaggregated these broad categories into the specific needs and priorities of different professional groups and how these might be delivered – for example, through short courses, other kinds of CPD and training that all enable good practice to be disseminated.

It might be argued that the use of 'management speak' terminology in some of the literature risks obscuring the essence of the message, that practitioners with good discipline-specific knowledge and understanding also need expertise in a range of strategic and operational implementation tools. This includes the ability to work with the other professionals and agencies that may be involved in the design and implementation of heritage conservation projects, and with the individuals and communities whose lives are affected by these proposals. Writing from an Australian perspective, Perkin (2010) suggests that such skills may be beneficial in managing a project to ensure that it is seen as representing the ambition of the community that is promoting it, rather than conforming to a predetermined model driven by external stakeholders.

This last point is significant; the increasing emphasis by public funding agencies such as the Heritage Lottery Fund on partnership and inclusion in the design, delivery and management of programmes linked to the conservation and reuse of heritage assets points to the targeting of a wider range of individuals. This is particularly important with regard to the building of capacity within a community to ensure the operational sustainability of schemes once the initial funding stream and physical works have terminated.

Conclusion

In essence, what we have discussed here is that capacity building is not only or usually a matter of developing technical and financial competence amongst volunteers, but recognising that, in the term 'capacity', we must include developing confidence — to ask the right questions, to know who to speak to, to know who to notify and so on — that is, to become the type of 'informed client' being nurtured by the Heritage Alliance's *Heritage 2020* initiative in its advocacy of better frameworks of advice and training for communities (2014: 6.11 and 6.14).

It is tempting to conclude that those in education and training, such as the authors, are increasingly in the business of producing prospective professionals with skills in 'fixing things'. This includes skills in operating within the framework of the law, ecclesiastical procedures and financial constraints and skills to fix things in terms of technical advice and good practice concerning design, maintenance and repair, project management, etc. But, increasingly, it is also to fix things in developing the capacity of those with whom they will work to have the confidence to engage with the Faculty system to achieve their aspirations for the parish and its community.

Notes

1 The faculty jurisdiction is the Church of England's system for regulating work to places of worship and a 'faculty' is a licence to carry out works of repair or alteration.
2 The 2015 Statements of Significance Creator website is at http://www.statementsofsignificance.org.uk/

References

Academy for Sustainable Communities 2007. *Mind the Skills Gap.* Leeds: Academy for Sustainable Communities
ACFSG (Archbishops' Council Faculty Simplification Group) 2012. *Report on Simplification of the Faculty Process* [online]. Church Care. Available at http://www.churchcare.co.uk/churches/faculty-rules-2015/simplifying-the-faculty-process [accessed 3 August 2015]
Centre for Rural Economy, Newcastle University 2014. *Reimagining the Rural: What's Missing in UK Rural Policy?* [online]. Newcastle University. Available at www.ncl.ac.uk/cre/news/NU%20CRE%20Rural%20Policy%20(web).pdf [accessed 3 August 2015]
Church Buildings Council 2015a. *Faculty Simplification* [online]. Church Care. Available at http://www.churchcare.co.uk/churches/faculty-rules-2015/simplifying-the-faculty-process [accessed 3 August 2015]

Church Buildings Council 2015b. *Open and Sustainable* [online]. Church Care. Available at www.churchcare.co.uk/churches/open-sustainable [accessed 3 August 2015]

Church Buildings Council 2015c. *Conservation Management Plans* [online]. Church Care. Available at http://www.churchcare.co.uk/churches/open-sustainable/guidance-documents-and-advice/conservation-management-plans [accessed 27 August 2015]

Church Care 2015. Guidance and advice [online] Church Care. Available at http://www.churchcare.co.uk/churches/guidance-advice [accessed 27 August 2015]

Crofts, S and Minnis, K. 2012. *Faith in Maintenance: Project Review*. London: SPAB

Dakers S. 2011. 'Church volunteering has little to do with the big society' [online]. *Guardian*. Available at www.theguardian.com/society/joepublic/2011/may/20/church-volunteers-big-society [accessed 3 August 2015]

Diocese of Exeter 2015. *Church Buildings* [online]. Diocese of Exeter. Available at www.exeter.anglican.org/church-life/church-buildings [accessed 3 August 2015]

English Heritage 2013. *Heritage Counts 2013*. London: English Heritage

FJR (Faculty Jurisdiction Rules) 2015 [online]. Available at http://www.legislation.gov.uk/uksi/2015/1568/schedule/1/made [accessed 3 August 2015]

Goddard, S. 2012. *Faith in Maintenance: Final Project Evaluation Report March 2007–January 2012.* Ipswich: Oakmere Solutions

Heritage Alliance 2014. *Heritage 2020* [online]. Heritage Alliance. Available at http://www.theheritagealliance.org.uk/tha-website/wp-content/uploads/2014/11/Heritage-2020-framework.pdf [accessed 3 August 2015]

Lukka, P., Locke, M. and Soteri-Procter, A. 2003. *Faith and Voluntary Action: Community Values and Resources*. London: Institute for Volunteering Research and School of Social Sciences, University of East London

Moore, S., Rydin Y. and Garcia B. 2015. 'Sustainable city education: the pedagogical challenge of mobile learning and situated knowledge'. *Area* 47(2), 141–9

Murray P. 2011. *The Sustainable Self*. London: Earthscan

ODPM. 2004. *Skills for Sustainable Communities (The Egan Review)*. London: ODPM

Perkin, C. 2010. 'Beyond the rhetoric: negotiating the politics and realizing the potential of community-driven heritage engagement'. *International Journal of Heritage Studies* 16(1–2), 107–22

Sayce, S. and Farren-Bradley, J. 2011. 'Educating built environment professional for stakeholder engagement'. In R. Rogerson, S. Sadler, A. Green and C. Wong (eds), *Sustainable Communities: Skills and Learning for Place-making*. Hatfield: University of Hertfordshire Press, 23–35

SPAB 2008. *The Good Maintenance Guide*. London: SPAB

SPAB 2015. *Faith in Maintenance* website [online]. SPAB. Available at www.spabfim.org.uk/index.php [accessed 3 August 2015]

19 The devil is in the detail

Capacity building conservation skills at the Stone Masons' Lodge

Sophie Norton

Introduction

The title of the conference that stimulated this chapter, Engaging Conservation, explicitly signifies an evolution towards a more inclusive practice of conservation. Earlier chapters in this volume have explored the background to this discursive shift in the heritage sector. The argument pioneered by Smith (2006), that there is a need to problematise and understand the competing values 'attributed to heritage', has arguably formed a new orthodoxy and underpins current approaches to identifying and managing heritage, which attempt to account for multiple, non-expert 'communal' interpretations. This paradigm is evident in both recent scholarship and academic critiques (Smith 2006; Fairclough *et al.* 2008; Avrami 2009), and in UK policy (English Heritage 2008; DCLG 2012). Taken together, they provide a theoretical rationale and several methodologies for understanding heritage as perceived by communities and other stakeholders, in addition to the 'authorised' expert.

Despite the notable developments that have taken place in the management of heritage in a broad sense, the Engaging Conservation project highlights that there are still barriers preventing people and communities from becoming involved in practical conservation. We are speaking here of material conservation practice as distinct from the umbrella discipline of 'heritage management', which can be seen to incorporate the identification, designation and presentation, as well as the conservation, of heritage. Conservation literature that has considered the subject (for example, Clavir 2009; Muñoz Viñas 2005) has pointed out the tensions that arise when attempting to physically conserve the social constructs of significance and authenticity through the treatment of objects, but has not been able to offer a sound resolution. Erica Avrami (2009) proposed an alternative approach in her examination of the process of conservation as itself a 'generator' of social value. This emphasises conservation activity as a catalyst of communal and social value, rather than a process to sustain or enhance it in the heritage, and potentially offers a new and genuinely dynamic way of interpreting the value of conservation, as opposed to the value of heritage.

This detailed case study of the Stone Masons' Lodge[1] is based on the discussions of three apprentices and explores how their particular group, which comprises a community of craftspeople in a conventional association with conservation, might interact and be affected by new forms of publicly engaged conservation. It demonstrates there are new opportunities for conservation practice to respond to the loss of intangible building crafts heritage, while building capacity in the sector and producing socio-economic value through skills development opportunities for apprentices. Conversely, it can also be seen that the holistic and more inclusive forms of heritage discourse identify significance that contrasts unexpectedly with the craftspeople's more traditional viewpoints. This leaves the craft community occupying the somewhat paradoxical position of expert without authority: it has been so inherently involved with heritage conservation through the twentieth century (Powys 1929: 4–7; Thurley 2013: 88) that its distinctive view as a central stakeholder in the detailed practice of conservation is overlooked, and the craft community's contribution remains accidental, opportune and tangential.

The methodology

As Regional Heritage Skills Coordinator at the University of York, the author of this chapter has worked with many people actively involved in conservation. Usually they are trainee craftspeople involved in practical building conservation to improve their skills for career purposes. Encounters with craftspeople often take place in quite formal settings, such as meetings or interviews, where individuals might give an account of their experiences of conservation. This chapter is based on a very different research encounter at a live conservation training project for three craft apprentices, which presented a rare opportunity to use participant observation research methods. The application of this methodology innovatively compares to previous baseline research into craft skills for conservation by the National Heritage Training Group (2013), which was prompted by skills shortages and addressed through interviews and surveys that explored the demand and supply of skills. Such a deductive approach overlooks a growing body of academic literature that studies the conservation crafts more creatively and without preconceived ideas about practitioner issues in the sector (Jones and Yarrow 2013; Marchand 2009).

Trevor Marchand (2009) is an anthropologist who adopts what he terms an 'apprentice-style field method', which involves immersing himself as an apprentice in the masonry and carpentry communities that he is studying. This has allowed him to 'learn about learning' through first-hand experience and he offers some rare insights into the training processes as perceived by the communities' actors. In an observational study of stone masons at Djenne World Heritage Site in Mali, he has found that there is communal value in the way that the teaching and learning of 'skilled performance and embodied practice [takes place] in a participatory forum located "on-site" ' (ibid.: 47). His descriptions of master mud masons producing a 'tiny, round aperture' could be interpreted as having an intangible value because it allows them to 'flaunt their skill and expertise' (ibid.: 56). These decorative features were the innovation of a master mason working in

Djenne in the mid-twentieth century and as such are not archaeologically accurate, conjectural, nor even required for adaptive reuse.

For much of the twentieth century, this innovative, creative craft practice was at odds with standards of fabric-based conservation in the UK, which developed through the nineteenth century as part of a broader European concern for preservation of historic monuments (Jokilehto 1999). In the century that followed, this trend became enshrined in important international policy (Athens Charter 1931; ICOMOS 1964) and national law, but the necessary acts of coalescence and refinement had an almost sterilising effect on the broader, more inclusive aspects of the early movement's concern for the value of creativity and handicraft practice (Ruskin 1849). The deficiencies that this caused were highlighted by the *Burra Charter* from 1999 (Australia ICOMOS 2013) which, through developing a broader value-based approach to interpreting heritage, showed that traditions outside the accepted fabric-based framework were being overlooked in conservation policy. International policy has developed in light of this; now the innovative and creative techniques of the stone masons at Djenne can be seen as in keeping with a tradition of crafts and craftsmanship, making the capacity to develop them an intangible heritage itself worthy of conservation.

It is also useful to refer here to the work of Sian Jones and Thomas Yarrow (2013), who have conducted ethnographic research in participant observation at Glasgow Cathedral. Their research explores the cathedral stone masons' inter-disciplinary relations with other conservation actors at the cathedral, as they negotiate the complex web of making decisions about repairing, replacing and recording stone. Through observing and interviewing a range of stakeholders at the cathedral, the authors conclude that authenticity 'can be accrued through the involvement of the masons, as embodiments of a craft tradition, who can carve something new, just as medieval masons would have' (ibid.: 17). They are not alone in finding this: Lindsey Asquith and Marcel Vellinga (2006) have made the case that vernacular buildings can be seen as just one part of a tradition that includes craft processes, which are as important as the material building, itself an empty physical manifestation of 'traditional building' in the active sense. Although Jones and Yarrow (2013; Jones 2010) do not find the distinction between tangible and intangible heritage useful, preferring to see the relationship as 'inalienable' instead, the researchers do agree that recognising multiple and holistic values is important in both heritage and conservation.

All three of the studies described above show conservation as a process of traditional building craft, supporting Avrami's (2009) proposition that conservation activity can and does generate additional value. In illuminating an under-appreciated, first-hand perspective of buildings crafts for conservation, they are able to highlight deficiencies in the positivist positions that base practical conservation decisions on either authorised or communal heritage discourses. They are able to do this because they adopt a truly qualitative methodology, aimed at finding out more about their subjects' first-hand human truth. When the author of this chapter observed the apprentices conserving the Stone Masons' Lodge – the case study for this chapter – it was similarly intended as a very open and inductive exploration of the value of the craft process to them, guided by a need to find out more about the heritage and social values of conservation. The research involved spending around two and a half days per week over a six-week period at the project, observing and working alongside the three apprentices, and recording observations in a fieldwork diary at the end of each day.

The building

The Stone Masons' Lodge is a nineteenth-century outbuilding located at the back end of a long, mediaeval burgage plot. Situated next to a footpath along a stream and visible from the road, a number of the building's characteristics contribute to the character and appearance of the area. It is traditionally constructed, with solid stone walls and a timber roofing structure that supports pan-tiles fixed to lathes with lime. These vernacular materials and features (including boarded windows and door) are typical of this part of rural England. Although it is neither scheduled nor listed, the building's heritage value is recognised in the local Conservation Area Character Appraisal. Such buildings 'provide an important break in the building line, afford an insight into how this area was previously used and valued and contribute to the sense of openness within the town' (RDC and NYMNPA n.d.). Importantly, it also enhances 'a lower scale element to the streetscape providing views of incremental ridge heights' and 'make[s] a positive impact on the Conservation Area' (ibid.).

Considered in relation to the four key value groupings identified in *Conservation Principles* (English Heritage 2008) – evidential, historical, aesthetic and communal – the values attached to the Stone Masons' Lodge most clearly derive from its evidential and aesthetic characteristics. Not only does it provide evidence about activity in the nineteenth century, a rectified photographic survey of the building has indicated that it was built in two phases (see Figure 19.1). Further, some of the larger stones in one of the phases are likely to have been taken from the mediaeval castle site across the road.

This evidential information is interesting but not unique, as similar interpretations can be inferred in other sites locally. Correspondingly, the building's aesthetic value, which relates to its appearance and the way it contributes to the character of this part of the Conservation Area, derives from several buildings' collective character. Although the appraisal does assert that 'these buildings should be conserved wherever possible' (RDC and NYMNPA n.d.), it is a pragmatic statement hinting at two divergent issues: finding conservation solutions for valued heritage is important but extremely challenging when that heritage is not rare, nationally significant or sustainably reusable. Situations like this are not unusual; according to the National Heritage Training Group (NHTG) around 94 per cent of pre-1919 buildings are Grade II listed or unlisted, which means that they are not registered on English Heritage's Heritage at Risk Survey and any neglect usually goes unrecorded and unimpeded. The condition of the building was one reason why it was selected as a 'live-site' training project for three

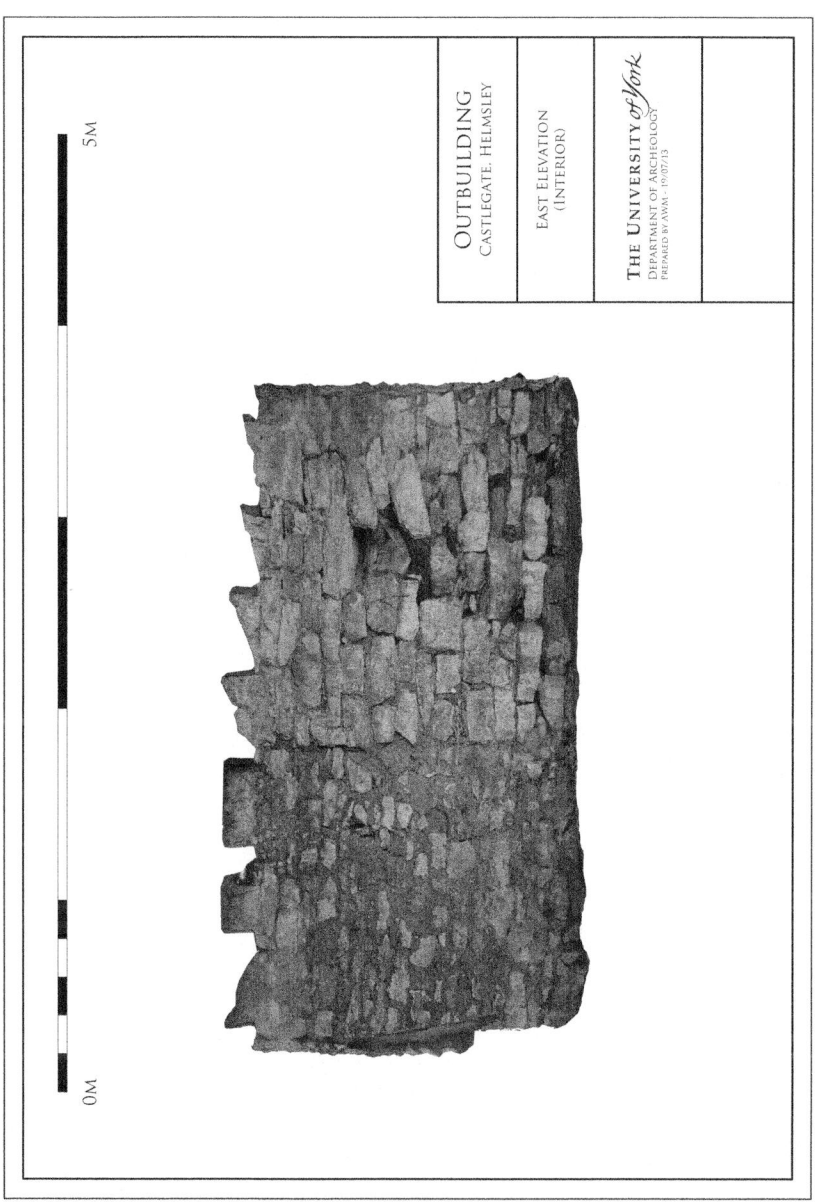

5M

0M

OUTBUILDING
CASTLEGATE, HELMSLEY

EAST ELEVATION
(INTERIOR)

THE UNIVERSITY of York
DEPARTMENT OF ARCHEOLOGY
PREPARED BY AWM - 19/07/13

Figure 19.1 Rectified photograph showing the two phases of the building, with some larger stones thought to be from the castle site nearby © Anthony Masinton, 2012

building craft apprentices. The project steering group also felt that the Stone Masons' Lodge's traditional construction methods but limited evidential value seemed to present an ideal opportunity for training.

The project

The project steering group was comprised of representatives from English Heritage, the Local Authority and four land-owning estates. For a period of two years, all of the estates had hosted a building craft apprentice who attended college and completed a nationally accredited qualification. In addition to the estate-based vocational training and the college-based underpinning knowledge, the steering group wanted the apprentices to engage in conservation in order to learn important theory and skills. To achieve this, several short courses and projects were planned where the apprentices could learn and practise a variety of skills, including using hot limes and conserving rural earthworks. One such project was the conservation of the Stone Masons' Lodge.

The project planning stage was conducted conventionally, with the owner and local planning authority appointing a conservation architect to draw up a simple plan and a guide specifications for repair to be discussed and re-evaluated on site, to form an important part of the apprentices' learning process (Figure 19.2). The building was then recorded and these documents were used for reference throughout the project. Arrangements for health and safety were made, and the works were then programmed on a Gantt chart so that the apprentices could set about conserving the Lodge under the supervision of a site foreman, a representative of building's owner and several independent craft and conservation specialists selected for their experience in training.

On the first day of the project, it became surprisingly clear how seemingly extraneous factors affected the work programme. Both the weather and the site culture, dominated by young people eager to learn, meant that we worked very quickly. The author's first fieldwork diary entry reads as follows:

> First day on site with the lads at the Lodge and I'm knackered! I've learnt an awful lot though. It's amazing how quickly the project plan and other paperwork is disregarded as soon as you get on site. We want to work on the roof this week, while the weather's good, so we need to change the dates that our specialist roofers come in. The truck broke down so we couldn't get rid of any waste.

It also became clear on the first day that there were several competing viewpoints about the repair and conservation techniques appropriate for the project: 'The architect wants us to retain the bow in the roof, but the [owners] "won't do that" and never do that. They also don't splice repair rafters. No wonder there's a lot of miscommunication … on building sites!'

Although the architect and Local Authority had agreed to this loose specification before the project began, miscommunication meant that neither party was

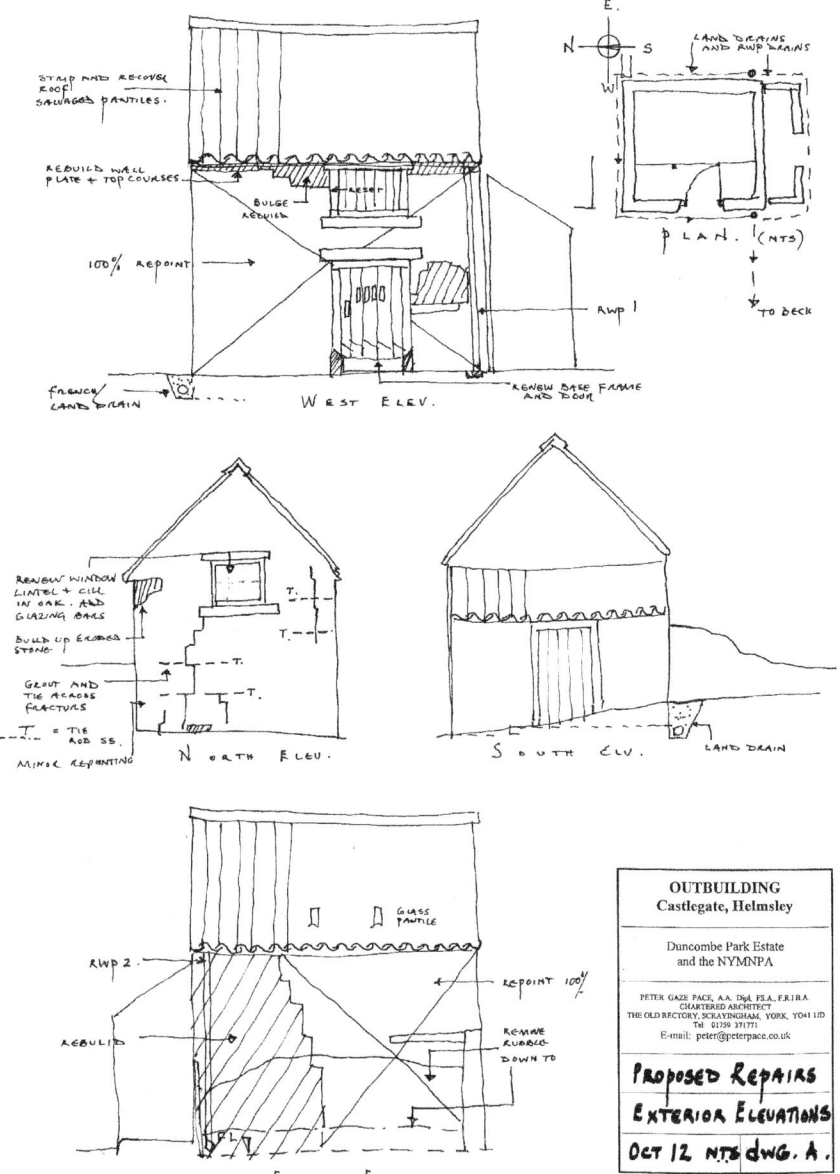

Figure 19.2 The apprentices did not give much attention to the architectural plans, elevations and specifications. © Peter Pace, reproduced by kind permission, 2013

able to attend site regularly, therefore their experiential knowledge of the building was not conveyed to the apprentices. As this was the type of knowledge that they respected and craved most, they and the steering group should have specifically requested a visit very early on. Instead, they generally referred to the site foreman when decisions about detail had to be made. Additionally, they seemed to be more comfortable exchanging ideas with the foreman (who quite literally spoke their language in that he employed similar terms for the building's features) as opposed to the architect and other professionals engaged in the project. This example from the author's fieldwork diary, which records a debate about timber replacement in the roof, interchanges the word 'rafter', which was used by the professionals and in all written and drawn communications, for 'spar' which was used verbally by the site foreman and apprentices: 'The [foreman] wanted to replace 8 of 10 spars, the ridge beam and the wall plate, but we've decided to retain the ridge beam, 4 rafters, repair 1 rafter and replace 5.'

This example demonstrates well how the detail of the architect's specifications, which said to 'replace rafters where necessary', were subject to negotiation and interpretation as the project progressed. Although this was not problematic for the steering group, which from the outset had wanted to involve the apprentices in oft-visited discussions about the nature of conservation repair methods, the apprentices themselves found it difficult that there was no single answer. Greater effort to engage in verbal discussions with professionals on site may have eased their frustrations, which are clearly demonstrated in the project evaluation forms that they completed: 'Too many people at some point on the same jobs so to[o] many opinions running around' and 'I understand what conservation means and what the architect wants. But no[t] sure how far you go before you replace with new.'

There was also tension in the way that the specifications were used on site. They were rarely referred to and quickly became scattered across the floor of the building. The apprentices said that: 'They could [have] been more [specific], and we ended up doing the methods differently to what he said' and '[They needed to be more detailed], clearer, [in] simpler language and shorter.'

Although one of the apprentices did say that they found the specifications useful, the author observed them making more use of the rectified photographs, produced before the project. These guided the rebuilding of part of a collapsed wall, and the reinstatement of the roof covering. In stark comparison to the specifications, the apprentices taped the photographs up inside the building for ease of use (Figure 19.3).

The apprentices' frustrations at debates about repair and replacement were frequently directed at the Local Authority-agreed specifications not being based on a thorough enough understanding of the place. Or, they might also have questioned the various specialist conservation craftspeople's decisions about trying to repair and retain as much fabric as possible. However, in reality, disagreements about methods of repair occurred between the site foreman and another builder representing the site's owner too. In the final week, there is a note in my fieldwork diary that states:

Figure 19.3 The apprentices pinned the rectified photographs up for ease of use © Author, 2013

> I called the lads to tell them to expect the scaffolders, they were pissed off because [the foreman] had been past and said that the guttering [that they had erected under the tutorage of the builder] looked stupid and had to be altered. There is some truth in this – because the eaves isn't straight, the guttering looked really far away from the eaves at one end, but it worked, which is the most important thing, surely!

Although I think the apprentices found this experience disheartening more than anything else, there is an element of disagreement underlying this apparently trivial issue, with one person valuing the aesthetics of the building and the other its functionality. Such divergences of opinion are increasingly endemic in a sector where value-led decisions based on multiple stakeholders' heritage have become the norm, which means achieving consensus can be difficult. Not only are craftspeople divorced from the heritage process of consultation and valorisation, which developed in the twentieth century and accounts for multiple, non-fabric-based perspectives, but there is also no corresponding framework or normative process for decisions about conservation. This means that the decisions about important detail made on site were often unilateral, conducted in a vacuum not permeated by the heritage process, though deeply embedded in practical conservation. The apprentices' disregard for the project paperwork, which itself reflects decisions they felt excluded from, symbolises how problematic and disenfranchising this position can be, and was in this case.

Conclusion

The steering group chose the Stone Masons' Lodge site for characteristics that were deemed appropriate for a training project. It was traditionally constructed, albeit with limited evidential value, in a neglected condition without being dangerous and, with limited options for reuse, it was unlikely to be conserved by any other means. However, when on site, the lack of evidential value meant that there was a great deal of debate around the need to expend extra time and effort on certain features. Common phrases were 'Well, it would be worth it if it was made of oak' and 'I would understand if it was a listed building'. On reflection, it seems that the apprentices' experiential knowledge of buildings and materials meant that they had preconceived ideas about what was valuable and what was not. While the building's aesthetic contribution to the conservation area could be referred to when its appearance risked being altered, through replacing features like the ridge board, the retention of unseen, poor quality and commonplace features was more difficult to justify. Although the apprentices acknowledged that some of the skills they were learning, such as splice repairing the rafters, were new, a lot of the treatment to the masonry was work they had done before.

The apprentices' evaluations corroborate the author's observations: while all said that they had learnt about project management and conservation, none thought they had learnt much about their specific craft occupation. A more complex building, which needed replacement structural and decorative elements in either stone or wood, might have challenged their craft skills more fully and engaged them in conservation efforts on a more challenging level. As it was, they found their experience of conservation confusing, frustrating and undemanding of their craft skills. The project would have undoubtedly been of greater value to them if they were able to debate and practise complex procedures and details within the framework of conservation. As semi-experienced craftspeople they wanted to be involved in difficult work to conserve the rare and high-quality buildings that they perceived as significant. In some ways, the limited evidential value of the Stone Masons' Lodge led to an increase in the amount of debate over the conservation solutions appropriate for the site, but it was a debate that the apprentices were reluctant to enter into because of their perception of its limited heritage value.

These observations offer insights into Avrami's (2009) suggestion about the process of conservation catalysing wider benefits: the nuances around the values individuals, and even communities, attach to both the object and process demand varied approaches. These apprentices were not fully satisfied by being involved in the conservation of the Stone Masons' Lodge, which they perhaps saw as 'conservation for conservation's sake'. They wanted to conserve sites that they deemed to have value, which, in relating to rarity, quality, age and craftsmanship, aligned with heritage assets in a more traditional sense. Finding that this craft community of study had a more traditional view of the heritage than the professionals who were steering the project was unexpected. Perhaps it was something that should have been anticipated given the relatively recent shift in the sector described earlier.

More significantly, given the focus of this volume, the apprentices valued certain aspects of the conservation process more highly. As craft apprentices trying to learn a skill, they were most interested in undertaking the challenging and complex details that conservation often demands. While this correlates with the findings of Jones and Yarrow (2013), Asquith and Vellinga (2006) and Marchand (2009), referred to above, it illustrates that the value of the conservation process can be subject to communal interpretation, as Smith has shown for heritage as a practice (Smith 2006). The idea that different communities might attach different values to the process of conservation, as detailed decisions are made on site, deserves further attention.

Finally, given the quite critical flavour of this chapter, I would like to reflect on the successes of the project. All of the apprentices were excited by the opportunity to take ownership of the Stone Masons' Lodge, referring to it as 'our building' on some paperwork. They enjoyed working together and being given the chance to 'get on' under their own steam. They said that they had learnt about project management and conservation principles, which are both essential elements of the conservation process, so in this respect the project achieved our objectives. Ultimately, and to quote Smith (2006) once more, the apprentices' engagement with the 'social and cultural process' of conservation was meaningful, and enabled them to 'understand and engage' with the heritage 'in the present'. That they have all gone on to gain further work within the sector, one with a specialist conservation firm, demonstrates the wider socio-economic benefits that projects building capacity in the conservation sector can produce.

Note

1 The building has been called the Stone Masons' Lodge here in order to anonymise the participant apprentices.

References

Asquith, L. and Vellinga, M. (eds) 2006. *Vernacular Architecture in the 21st Century: Theory, Education and Practice*. London: Taylor & Francis

Athens Charter 1931. *The Athens Charter for the Restoration of Historic Monuments*. ICOMOS [online]. Available at http://www.icomos.org/en/charters-and-texts/179-arti-cles-en-francais/ressources/charters-and-standards/167-the-athens-charter-for-the-restoration-of-historic-monuments [accessed 24 August 2012]

Australia ICOMOS 2013. *The Burra Charter: The Australia ICOMOS Charter for Places of Cultural Significance* [online]. Available at http://australia.icomos.org/wp-content/uploads/The-Burra-Charter-2013-Adopted-31.10.2013.pdf [accessed 12 August 2015]

Avrami, E. 2009. 'Heritage values and sustainability'. In A. Richmond and A. Bracker (eds), *Conservation: Principles, Dilemmas, and Uncomfortable Truths*. Oxford: Butterworth-Heinemann, 177–83

Clavir, M. 2009. 'Conservation and cultural significance'. In A. Richmond and A. Bracker (eds), *Conservation: Principles, Dilemmas, and Uncomfortable Truths*. Oxford: Butterworth-Heinemann, 139–49

DCLG (Department of Communities and Local Government) 2012. *National Planning Policy Framework*. London: Stationery Office

English Heritage 2008. *Conservation Principles, Policies and Guidance for the Sustainable Management of the Historic Environment*. London: English Heritage

Fairclough, G., Harrison, R., Jameson, J.H. and Schofield, J. 2008. *The Heritage Reader*. London: Routledge

ICOMOS 1964. *The International Charter for the Conservation and Restoration of Monuments and Sites (The Venice Charter 1964)*. Paris: ICOMOS [online]. Available at http://www.icomos.org/charters/venice_e.pdf [accessed 24 August 2015]

Jokilehto, J. 1999. *A History of Architectural Conservation*. Oxford: Butterworth

Jones, S. 2010. 'Negotiating authentic objects and authentic selves'. *Journal of Material Culture* 15(2), 181–203

Jones, S. and Yarrow, T. 2013. 'Crafting authenticity: an ethnography of conservation practice'. *Journal of Material Culture* 18(3), 3–26

Marchand, T.H.J. 2009. *The Masons of Djenne*. Bloomington: Indiana University Press

Muñoz Viñas, S. 2005. *Contemporary Theory of Conservation*. Oxford: Butterworth Heinemann

National Heritage Training Group 2013. *Traditional Building Craft Skills: Reassessing the Need, Addressing the Issues*. London: NHTG

Powys, A.R. 1929. *The Repair of Ancient Buildings*. London: SPAB

Ruskin, J. 1849. *The Seven Lamps of Architecture*. London: Smith Elder

RDC and NYMNPA (Ryedale District Council and North York Moors National Park Authority) n.d. 'Conservation Area Character Appraisal'. Unpublished

Smith, L. 2006. *Uses of Heritage*. London: Routledge

Thurley, S. 2013. *Men from the Ministry: How Britain Saved its Heritage*. London and New Havne, CT: Yale University Press

Index

Page numbers in *italic* refer to illustrations, figures and tables. Added to a page number 'n' denotes notes.

Printed in Great Britain
by Amazon